CAMBRIDGE BIOLOGICAL STUDIES

GENERAL EDITOR
C. H. WADDINGTON

THE EPIGENETICS OF BIRDS

THE EPIGENETICS OF BIRDS

BY

C. H. WADDINGTON
Sc.D., F.R.S.

*Professor of Animal Genetics in the
University of Edinburgh*

'EPIGENETICS'. The science concerned
with the causal analysis of development

CAMBRIDGE
AT THE UNIVERSITY PRESS
1952

CAMBRIDGE
UNIVERSITY PRESS

University Printing House, Cambridge CB2 8BS, United Kingdom

Cambridge University Press is part of the University of Cambridge.

It furthers the University's mission by disseminating knowledge in the pursuit of education, learning and research at the highest international levels of excellence.

www.cambridge.org
Information on this title: www.cambridge.org/9781107440470

© Cambridge University Press 1952

First published 1952
First paperback edition 2014

A catalogue record for this publication is available from the British Library

ISBN 978-1-107-44047-0 Paperback

PREFACE

Since the very beginning of embryology, the chick embryo has perhaps been more studied than any other, and an almost overwhelming number of books and papers have been written about it. The older literature has been summarised in the monograph of Lillie (1919). Nevertheless, the preparation of still one further book on the subject at this date does not seem to call for an apology; rather, one feels impelled to point to the recent war as an excuse to explain why it did not appear ten years ago. The last quarter of a century has seen a revolution, not only in our knowledge of the anatomy and physiology of the early stages of avian development when the fundamental ground-plan of the animal is being laid down, but also in the general framework of embryological understanding against which the particular problems of the chick are to be seen.

The modern movement in embryology, beginning perhaps with His in 1874, reached its first fruition, at least as far as vertebrates are concerned, with the work of Spemann and his school in the years around 1920. In broadest outline, the new point of view stemmed from two types of technical advance. First, from a logical if not from a chronological standpoint, the application by Vogt of the method of marking groups of cells by vital dyes revealed the extent and importance of morphogenetic movements which take the form of tissue streams, thus opening our eyes to the limitations of the older techniques of comparing a sequence of stages, each fixed as an instantaneous cross-section of the continuous process of development. Secondly, the elaboration by Spemann of more powerful operative techniques began for the first time to reveal, not merely new phenomena for which an explanation must be sought, but at least the beginning of a causal analysis of the physiological processes on which embryogenesis depends. Embryology, from this time, ceased to be a purely descriptive science, fitfully illuminated by evolutionary speculations. In future no account of our embryological knowledge can be considered well balanced unless it includes an adequate discussion of our growing understanding of the dynamic and causal aspects of the phenomena; and it is in these fields, rather than in the more classical descriptive phases, that the growing points of the science are to be found.

Very soon after the first successes of Vogt and Spemann, attempts were made to apply their methods to the chick. Gräper, using cinematography, and Wetzel, using vital-stain marks, began in the late 1920's to discover the real events concerned in the origin of the germ layers and of the primitive streak, rescuing these phenomena—the fundamentals of avian development—from the unduly imaginative grasp of such phylogenetic speculations as the Concrescence Theory. Although the techniques are not quite so simple or satisfactory as in the amphibians, vital-staining methods, particularly in the hands of Pasteels, have immensely deepened our understanding not only of the chick but of several related forms: the teleosts, somewhat removed phylogenetically but similar in the meroblastic cleavage of the egg, and the reptiles, the nearest evolutionary relatives of the birds.

Techniques for operative experimentation on the bird embryo were also rapidly developed. Simple division of the embryo within the shell has been practised almost as long as the rather similar separation of blastomeres in the amphibians. Defect experiments, in which small areas of the blastoderm are killed by heat, electrolysis or localised X-irradiation, have also been made for many years. The blastoderm *in ovo* is under some tension, and there is a strong tendency for it to give way at the site of injury, so that the wound expands into a large hole, which may greatly distort the embryo and even render it difficult to identify the parts which develop. Nevertheless, some workers have succeeded in producing useful results in this way, although the deceptiveness of similar experiments in the amphibians has taught us the need of great caution in interpreting them. A slightly more penetrating method is that of making a graft on to the chorio-allantois (or in later developments into the coelom of an older embryo) from which it derives a blood supply which enables it to develop. The technique has been particularly used by American authors, beginning with Hoadley (1926 *et seq.*) and continuing with Willier and his pupils. It has certain technical defects, which will be discussed later (p. 3), but its main limitation is that it can only reveal the 'potentialities' of the isolated fragment. These are to be regarded rather as problems than as explanations. Causal relations can be discovered only when we can investigate the interactions between two entities, as was done by Spemann when he described the results of the action of a graft on the surrounding host tissues. The first results of this kind in the avian embryo were obtained by Waddington (1930) who used the method of *in vitro* cultivation to keep whole embryos alive in a situation in which the necessary manipulations were possible. Since this beginning, a considerable body of data has been collected, largely paralleling the work which has been performed by the Spemann school on the amphibians.

The newer descriptive and experimental studies on the chick have been only rather cursorily summarised (Waddington, 1939, 1950; Rudnick, 1944, 1948) and have as yet found their way only in part into monographs of experimental embryology such as those of Weiss (1935), Spemann (1936), Dalcq (1938, 1941) and Lehmann (1945). This may be partly because no one of these authors has himself worked on this form. In part it may also have been a result of a lack of coherence in the literature which seems particularly to have afflicted this field, which has tended to be a stamping ground for several distinct groups of workers, each of which built up a body of doctrine of its own, and sometimes conducted vigorous polemics within itself, without apparently paying much attention to work going on in other parts of the world. Thus there was a German group, consisting mainly of Wetzel, Gräper, Kopsch and Holmdahl, joined later by Pasteels, who were concerned mainly with the kinematics of gastrulation and embryo formation. The American group under the inspiration of Willier continued for many years to make experiments involving the isolation of complex fragments containing ectoderm, mesoderm and endoderm, usually on the chorio-allantoic membrane. In England, Waddington and a few others performed operations which made it easier to bring their results into line with the phenomena which had been revealed by Spemann in the amphibians and were by then well known. The work of all these three groups had reached a certain degree of finality by the outbreak of the war; nevertheless, their results have not yet been critically reviewed in relation to each other. Since then, the European work, which suffered the greatest interruption, has hardly got back into its stride. American embryologists, however, have been carrying forward their interests into many new and exciting fields. One may mention the applications of biochemical and biophysical techniques to problems of avian morphogenesis, with such important contributions as those of Jacobson and Brachet from Europe, and Moog, Hobson, and others in America. This has, indeed, been the main centre of interest in the last few years as far as the early stages of morphogenesis are concerned. The other main fields of recent effort have dealt with later stages of organogenesis. One may instance the work of Hamburger, Saunders and others on the limb buds, of Hamburger and his collaborators on the central nervous system, of Landauer and his group on certain genetic types, and of the Lillie and Willier schools on the development of the plumage and its pigmentation.

The central purpose of this book has been to review the literature, mainly dating from the 1930's, which deals with the primary stages of morphogenesis. Work on later stages, much of it published more recently, is reviewed rather more cursorily. It has, of course, been extremely difficult to decide where to stop. There is an immense literature on tissue culture studies of chick embryonic tissues and cells; in most cases the donor embryos had completed their basic morphogenesis and much of their histogenesis, so that the problems at issue are somewhat different, though of course related, to those of what may now be called classical experimental embryology. This material has been omitted almost completely. Again, the experimental study of sex determination and differentiation is an enormous but somewhat specialised field, which is probably better treated from a comparative point of view which does not confine itself to one Order of vertebrates; and for this reason, and because it would unduly expand the scope of the book, it has also been excluded. In fact, it may perhaps be said that the subject of pigmentation and plumage development has been taken as the borderline where experimental embryology shades off into other fields of experimental biology. A chapter is devoted to it, but the treatment is less thorough, and the coverage of the literature less complete, than that which has been given to the more central themes.

I wish to express my deep indebtedness to Mr A. D. Roberts for the care he has given to the exacting task of preparing the illustrations, and the great skill he has shown.

C. H. W.

EDINBURGH
August 1951

CONTENTS

CONTENTS xi

LIST OF TEXT-FIGURES

CHAPTER I

TECHNICAL METHODS

INVESTIGATION of the epigenetics of the chick has been considerably impeded by the technical difficulties which have to be overcome. Some of these are inherent in the nature of the embryo itself. The tissues of the blastoderm are more adhesive to one another than those of other embryos, such as the amphibian, which have proved most suited to the operative procedures of experimental embryology. Moreover, the cells are comparatively very small. Thus it is not, and probably will never be, possible to separate different presumptive areas in the chick with the precision which is a matter of course in the Amphibia. It is particularly unfortunate that the separation of the mesoderm from the ectoderm in the gastrulation stage is almost impossible to achieve.

Other difficulties arise from the nature of the situation in which the embryo is found. It lies within the egg-shell, covered by a layer of viscid albumen and a relatively tough vitelline membrane, while below it is the mass of semi-fluid yolk. Many experimenters, from the middle of the last century onwards (Valentin, 1851), have overcome these obstacles and made operative attacks on the embryo in its natural situation. (Beguelin (1751), was perhaps the first to keep the embryo alive and under observation after opening the shell, but without attempting any experimental interference with it.) An outflow of yolk through any holes which may be made in the vitelline membrane and blastoderm is a hazard which frequently disturbs the later development of such specimens. If this can be kept within bounds, the embryos may be reared in the shell for periods up to a week or more. The technical problems involved have been discussed in some detail by Wolff (1936). In his view the most serious difficulties arise during the first two days of incubation, during which the embryo is normally rather deeply buried by albumen, and runs great danger of dying through desiccation if this covering is removed in order to operate on it; in later stages, the undisturbed embryo tends to float towards the surface of the gradually liquefying albumen, and the main danger in operated specimens is of

WEB   I asphyxiation. Wolff recommends always withdrawing a few cubic centimetres of albumen from the pointed end of the egg before opening a window in the shell over the blastoderm. After the operation, the window is sealed with a cover-glass affixed with wax. If the embryo is of more than 36 hours' incubation, this can be done without replacing the albumen, and the egg can be incubated with the window upwards, when the blastoderm will lie immediately beneath it; but in younger specimens it is advisable to inject the albumen again through the hole in the pointed end, and to turn the window to the side, so that the blastoderm lies under intact shell at least until the second day, when the egg may be turned again to bring it beneath the window. These precautions, however, are not necessary unless it is desired to cultivate the embryos until a very late stage. Many of the most important studies made on embryos *in ovo* (such as the vital-staining experiments of Wetzel and Pasteels, or the cinematographic studies of Gräper) the essential observations were made over a shorter period and the young blastoderm was left beneath the window the whole time. Observation, both at the time of the operation and also later, may be facilitated by a light vital staining of the embryo; neutral red has usually been employed for this purpose.

The difficulties of operating on the blastoderm *in ovo* have led to the search for methods of keeping alive embryos which have been removed and operated on in more favourable circumstances. The first of these methods involved the grafting of blastoderms, or fragments of them, into subcutaneous sites in adult birds (Féré, 1895; Féré & Elias, 1897). Later workers have varied this method by making the grafts into newly hatched chicks (Strangeways & Fell, 1926). Much more extensive use had been made of grafts on to the chorio-allantoic membrane of chick embryos of some 7–9 days' incubation, a method introduced by Rous & Murphy (1911) for the investigation of tumours and first used in connection with embryology by Dantchakoff in 1917 (cited by Dantchakoff, 1924). A very good percentage of 'takes' can be obtained on this implantation site, and histological differentiation is often surprisingly normal. It will be seen later, however, that even from a histological point of view, the site must be considered in some respects inhibitory, since some tissues (in particular neural tissue) which may develop from a given fragment *in vitro* may fail to appear in a similar chorio-allantoic graft. There is also the objection, perhaps purely theoretical, that the blood in the chorio-allantoic circulation contains substances capable of causing inductions in the amphibian embryo, and the conditions cannot, therefore, be accepted *a priori* as neutral (Waddington, 1935); but there is no evidence that difficulties of the kind to be expected on these grounds ever arise in practice. A more considerable limitation is the fact that both growth (Hoadley, 1929) and, in particular, morphogenesis are greatly impeded. The graft becomes surrounded by loose mesenchymatous tissue from the host, and usually forms a more or less spherical mass within which the different histological types of tissue developing from the graft lie in chaotic disorder. Such morphological disturbance may, of course, prevent the appearance of certain histological types of differentiation, if the latter depend on inductive reactions between tissues which require to be brought together by the normal morphogenetic processes.

The technique used by the American experimentalists for carrying out the grafting was systematically worked out by Willier (1924) and is fully described in Hamburger's valuable *Manual* (1942 *b*).

Grafts have also been made to other embryonic sites. Probably none of these surpasses the chorio-allantois in its capacity to support histogenesis, but some of them are more favourable to morphogenesis. This applies in particular to transplantation to the coelom near the lateral margin of the body of a third-day embryo. The method was introduced by Hamburger (1938) and is fully described in the *Manual* mentioned above. Remarkably complete morphological development is obtained, particularly of isolated organs such as limbs, eyes, etc., even from quite early stages before their rudiments can be recognised (cf. Rudnick, 1945 *b*).

Good morphogenesis may also occur if fragments of the embryo are explanted into tissue cultures of the appropriate kind. Explants into small quantities of plasma-extract medium, such as are used in the well-known hanging-drop cultures, usually tend to break down into separate cells which wander out into the medium, and in such circumstances histogenesis is suppressed except in the most central part of the fragment, while morphogenesis is, of course, almost entirely prevented. Such cultures, however, have been employed to give valuable insight into some of the causal factors at work during development. More normal development may be obtained if larger fragments are used, and the quantity of medium increased, as in the watch-glass method of Strangeways & Fell (1926) and Fell & Robison (1929). The zone of outgrowth is relatively much smaller in such cultures, while the central portion is large enough for considerable differentiation to take place. Some examples of the results achieved will be described later (p. 165). It must be confessed, however, that satisfactory differentiation rarely occurs from fragments isolated before the appearance of well marked rudiments, and in these the main processes of determination are usually complete; thus the technique has proved applicable only to the study of the later details of development and not to the critical initial stages.

The great advance of our understanding of epigenetics which has been made in recent years has, in almost all groups of animals, depended on the discovery of methods which make it possible to investigate experimentally the mutual interactions between different parts of the embryo. This can only be adequately done when, in the still labile germ, fragments can be transplanted from one situation to another. Such experiments are extremely difficult in chick embryos *in ovo*, though fairly recently Cairns (1937) and Rudnick (1944) have been stimulated by the success of the *in vitro* method, shortly to be described, to repeat some of its results *in vivo*. The various isolation and grafting methods are also inadequate for this purpose, since the host embryos have long passed the period of primary determination and can no longer exhibit the kind of reaction to the graft in which we are interested; while the mechanical conditions within the grafted masses make it difficult or impossible to study the interaction between two embryonic fragments of the relevant age. Successful experiments of this kind were in fact first made in the chick by an adaptation of the tissue culture technique. This method was used, not in the usual manner to test the potencies of an isolated fragment, but simply to keep the whole blastoderm alive in a situation in which it is more accessible to operative procedures than it is in its normal situation inside the egg-shell. By far the greater part of our knowledge of the epigenetic reactions proceeding within the early embryo has been derived from the use of this technique, which renders the avian blastoderm only slightly less approachable than the amphibian gastrula. The method has more recently been shown, by Spratt, to open up still other avenues of inquiry.

In vitro *cultivation*

The first attempts to cultivate the chick embryo *in vitro* were made by McWhorter & Whipple (1912), who used the hanging-drop technique. They obtained no differentiation from embryos explanted before the formation of the head process, and their best results were with embryos of 10–12 pairs of somites, one of which lived 31 hours. Sabin (1919) used the method to observe the formation of blood vessels; the average life of her cultures was about 5 hours. A. Brachet (1912, 1913) and Maximov (1925) who attempted to cultivate young rabbit blastocysts *in vitro* also had rather modest success. It was not till the 'watch-glass technique' for the culture of large explants was developed at the Strangeways laboratory that any noteworthy success was attained. Some time in the late 1920's, D. H. and T. S. P. Strangeways made some preliminary attempts to cultivate whole blastoderms and obtained promising results. Their lead was followed by Wadding-ton in 1929, under the guidance of H. B. Fell, whose description of the method (Fell & Robinson, 1929) is still the *locus classicus* for this technical innovation. In essentials it consists in no more than the explantation of the blastoderm (or other largish fragment) to the surface of a clot composed of equal parts of fowl blood plasma and embryo extract, which is contained in a watch-glass enclosed in a small damp chamber; the latter is usually provided by a Petri dish containing a ring of moist cotton-wool—the essential point is that the surface of the culture must never be allowed to become dry. Details of the technical manipulations special to work with chick blastoderms will be found in Waddington (1932). Cultures can also be made in other types of vessel, e.g. in one embryological watch-glass, to which a second is sealed with paraffin (Rudnick, 1938a; Spratt, 1940). The only essential points seem to be that the clot should be fairly large, its surface moist, and possibly that a large volume of air should be available to the embryo.

By this technique, embryos can be kept alive for some 2–3 days. The development which they undergo during this time appears to follow very closely the normal course. There is, however, a general tendency for ectoderm to form cysts *in vitro*, and these are usually found in the outer parts of the area vasculosa of young embryos. As will be described later, such formation of cysts also affects the ectoderm of the area pellucida when it is cultivated after removal of the endoderm.

The rates both of growth and of differentiation are slower *in vitro* than *in vivo*. No exact study of the alterations of rate has been attempted but certain remarks can be made from a consideration of material which has accumulated. According to Lillie's (1919) chart, 1·04 hours are required, on the average, for the formation of one somite under normal conditions, between the stages of 2 and 27 somites. The same figure, for 34 specimens grown *in vitro*, is 1·65 with a variation between 2·44 and 1·15. Both figures apply to incubation at approximately 37·5° C. For accurate results more carefully controlled experiments would be necessary. Nevertheless, the ratio of about 1 : 1·5 between the *in vivo* and *in vitro* differentiation rates, can probably be taken as approximately correct. There is no evidence that this slowing of the rate of development affects the young stages more than the old or vice versa; for example, a blastoderm transplanted after 2 hours' incubation had 20 somites at the end of 72 hours, a stage which would be reached by a normal embryo in about 43 hours, giving a ratio of about 1 : 1·7.

The rate of growth (increase in length) is very variable *in vitro*, but is probably always considerably slower than *in vivo*. The stage of differentiation attained by an embryo is very largely independent of the absolute size, and, if a blastoderm is explanted at an early age and thus is affected by the slowing of the growth rate for the greater part of its life, embryos may be formed which, compared with the normal, are very much too small for their stages of differentiation. Thus an embryo cultivated for 72 hours after a preliminary incubation for 2 hours, has some 18–20 somites, although its length is only about 2 mm., whereas the normal length for such a stage should be about 5·3 mm. (from Lillie's table). It cannot be decided, with the material at hand, whether the slowing of the growth rate is greater if the blastoderm is transplanted at an early stage, or whether the marked effects seen in such embryos is entirely due to the fact that the slower rate of growth has been in action for a greater proportion of the total life.

It is clear from these facts that the processes of growth and differentiation can be, at least to some extent, dissociated from one another, so that a given stage of differentiation may be reached without the normal size being attained. Needham (1933) has discussed the evidence for such dissociability in other materials. It is, however, perhaps unjustified to assume that the formation of a normally shaped embryo proceeds by exactly the same route as it would do *in vivo*; we may be dealing with a paragenetic, rather than a normogenetic, teleogenesis. This doubt is particularly insistent in relation to the morphogenetic movements of the primitive streak stage. Our general knowledge of the regulatory abilities of vertebrate gastrulae would certainly not lead us to rule out the possibility that a normal end-result might be attained even if, owing to some impediment to the normal expansion of the blastoderm, the morphogenetic movements were forced to take a somewhat abnormal course. There is no direct evidence that this is so; but the possibility that it may be so remains as a doubt which may have an important bearing on investigations of gastrulation movements conducted on cultured embryos, such as those of Spratt (1947 *a, b, c*).

Modifications in the culture technique have recently been introduced by Spratt (1947 *d, e*). He first showed that the culture medium could be considerably simplified, development being possible on a clot stiffened by agar and containing diluted egg albumen or yolk; the former has the advantage that it possesses some bacteriostatic powers, and makes it unnecessary to observe fully rigorous precautions to ensure sterility. On the other hand, it is doubtful whether development proceeds so far in such media. Spratt appears rarely to have observed embryos in which regular circulation of the blood was established. Taylor & Schechtman (1949), using similar methods, claim that their embryos die at about the time when blood circulation ought to start, owing to failure to establish active circulation between the intra- and extra-embryonic vessels. However, this can hardly be the whole story, since embryos without proper circulation (e.g. after removal of the heart, p. 180) can survive for the order of a day on the plasma-extract medium. On that medium, also, the establishment of the circulation is a critical phase, but quite a large proportion of embryos succeed in passing through it and developing a regular flow of blood. Even on that medium, however, they do not maintain this for more than about 24 hours, the circulation eventually coming to a standstill, apparently owing to the coagulation of the blood in the vessels. The embryo is by that time too thick to obtain its nutrition and aeration by diffusion, and necrosis soon follows, starting at the lower surface.

It seems probable that the nutritive materials are less accessible

8 TECHNICAL METHODS

to the embryo on an agar base than they are when the clot is formed of coagulated plasma. The latter is certainly digested and liquefied fairly rapidly, presumably by extra-cellular enzymes secreted by the blastoderm. On agar, however, development does not appear (in preliminary experiments of Waddington) to take place satisfactorily unless the clot is initially very soft: Spratt also recommends a very low concentration of agar (0·25 %). The softness of such clots is rather a disadvantage when it is desired to operate on the explanted embryos, since their surface easily becomes broken up with furrows into which folds of the blastoderm sink.

Spratt and others have used the method of *in vitro* cultivation to study the nutritive requirements of the chick embryo during the period up to the stage of about 12 somites. Spratt (1948*a, b*) after proving the adequacy of White's (1946) complete medium, proceeded to show that the only essential constituent of this is glucose. Mannose was equally efficacious, and fructose, galactose and maltose could also support differentiation, while lactose and sucrose could not. Needham & Nowinski (1937) had previously shown by biochemical methods that the early embryo can glycolyse mannose nearly as well as glucose. Taylor & Schectman (1949), however, showed that the effect of yolk extracts on development is not solely due to their content of free glucose, since there is a non-dialysable heat-stable factor which has a limited effectiveness.

A new series of 'normal stages' in the development of the chick, from the beginning of incubation to the time of hatching, has recently been described, and illustrated with figures, by Hamburger & Hamilton (1951) but has of course not yet had time to come into general use.

CHAPTER II

THE EARLY DEVELOPMENT

First stages and the formation of endoderm

The avian egg is difficult of access before it is laid, and although considerable effort has been devoted to the study of the growth of the oocyte, fertilisation and cleavage, scarcely anything has yet been discovered which links these processes directly with the epigenetic events to which the later formation of the embryo is due. This phase of development can, therefore, be passed over rapidly in this book. The most recent reviews of the structure and growth of the oocyte are those of Conrad & Scott (1938) and Romanoff & Romanoff (1949). In these events, the polar structure of the ovum is, of course, very marked, the germinal vesicle moving from its originally central position so that it comes to lie immediately under the periphery on one side; the course of its migration is indicated by the column of white yolk, known as the neck of the latebra, which leads from the centrally placed latebra proper to the wider area of white yolk ('Pander's nucleus') immediately under the vesicle. The polar structure is clearly related to the position of the ovum in the follicle, the germinal vesicle lying directly under the follicle stalk. The region of the germinal vesicle can be compared to the animal pole of other less yolky eggs. In birds, it is the region from which the entire embryo and associated structures will originate, and is distinguished by containing not only the egg-nucleus but also the greater part of the cytoplasm, which forms a lens-shaped thickening around the germinal vesicle and, apart from that, extends as only an exceedingly thin peripheral sheet around the yolk, immediately under the vitelline membrane.

The existence of a polar axis provides less than the minimum structure from which the development of an embryo could be derived. We require also another axis at right angles, which would define a plane of bilateral symmetry. No indications of such an axis appear to have been described in birds. As we shall see (p. 65), the plane of bilateral symmetry, which is later marked by the formation of the primitive streak within it, seems to be very labile in the early stages of avian development, and does not become firmly determined until the streak actually appears. It is, therefore, perhaps unlikely that any clear indication of it will ever be found in the unfertilised stage; the bilaterality may develop more slowly and in a less clear-cut manner than it does even in the Amphibia, where also it shows some degree of lability (cf. Lehmann, 1945).

Fertilisation takes place in the oviduct. Many sperm enter the blastodisc surrounding the germinal vesicle. The supernumeraries, which do not unite with the egg-nucleus, may play some part in the early stages of the digestion of the yolk, but soon degenerate. The early cleavages have been described by several authors (e.g. Olsen, 1942) but nothing definite is known as to the relation of the cleavage planes to the later symmetry.

We have almost no experimental evidence concerning the events proceeding at these early stages, but Sturkie (1946) has recently shown that if hens are cooled (to about 36·7–38·3° C.) before the laying of an egg, up to 8 % of the eggs produced exhibit twinning of the axis or other abnormalities, and it seems certain that in many of these cases the cooling occurred before the first cleavage of the egg. The mechanism of the effect remains completely obscure.

At the time of laying, the blastoderm of the chick is a multi-cellular circular disc. This will shortly become differentiated into two distinct layers, lying one on top of the other; in fact, the separation is usually already under way when the egg is laid, and is completed wit hin the first few hours of incubation. These two layers are perhaps best referred to by the non-committal names of epiblast for the upper and hypoblast for the lower. As will be seen in more detail later, the upper layer certainly gives rise both to ectoderm and mesoderm and some authors claim that some endoderm may also arise from it in normal development. It has sometimes been referred to as the 'mesectoderm', a word which would not only conceal the last contribution referred to, but would also tend to encourage a confusion with the material which passes into the mesoderm from the neural crest (and which has also been called mesectoderm, although better termed ecto-mesoderm). The hypoblast is in some ways a less confusing tissue, since there is little question but that it develops only into tissues which can be considered as endoderm, and it appears quite justifiable to refer to it as 'hypoblast' or 'endoderm' indifferently.

There is, however, a possibility, which will be discussed later (p. 17), that a further contribution to the endoderm is made at a later stage by cells originating in the primitive streak.

The processes which lead to the formation of a double-layered blastoderm have been extensively discussed since very early times, but they are still, perhaps, not very convincingly understood. At

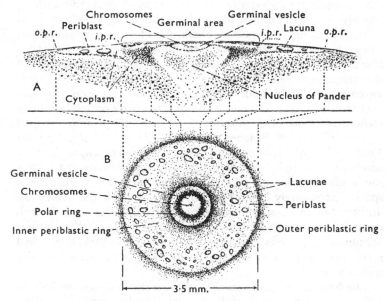

Fig. 1. The structure of the blastodisc in the unfertilised egg. Vertical section above, and surface view below. (From Romanoff & Romanoff.)

the beginning of the process, i.e. at or just before the time of laying the blastoderm is in the form of a fairly solid layer, the tissue being more coherent on the surface, while in its lower parts the cells are less closely connected to one another. Beneath the whole mass is a space which usually appears as a cavity in sections, but which in life is full of liquefied yolk; this is the subgerminal cavity. One of the earliest views on endoderm formation was that of Balfour (1873), who suggested that the lower layer is produced by the coalescence of the looser under-parts of the blastoderm to form a coherent sheet roofing over the subgerminal cavity. The most recent views have in most respects returned to this conception.

Several other hypotheses, however, have been advanced with more or less conviction. Perhaps the most famous of these is the suggestion that at the posterior margin the germ wall (i.e. the edge of the blastoderm, a locality in which the lower cells are intimately in contact with undigested yolk) becomes folded under and takes part in a true invagination. This suggestion, of which the most important supporter in earlier times was Duval (1884, 1888), was given a fresh lease of life by Patterson in 1909, who figured sections through the posterior margin of pigeon blastoderms. These showed a thickening of the cellular region, which was very sharply marked off from the nearby yolk; and Patterson claimed that the appearances showed that tisue was actually being invaginated round the blastoderm edge. This view has been generally accepted until recently, and is still quoted with approval in many standard text-books of embryology, even as lately as in the revised second edition of Brachet's valuable *Traité d'Embryologie des Vertébrés* (1935). However, it has now been shown, with some conclusiveness, to be erroneous. Pasteels (1937 b) repeated Pattersons's observations on pigeons and did indeed find that some specimens strongly resembled those figured by the American author; but this was by no means always the case, and Pasteels points out that, even in those which Patterson figures, the pictures are not unambiguous in their implications, and cannot be accepted as convincing proof of the movements which Patterson deduces from them.

The first of the more recent studies on the details of endoderm formation is that of Mehrbach (1935), who, while accepting Patterson's posterior invagination as a partial source of the hypoblast, suggested that a further important contribution was made by individual cells being set free from the bases of numerous small folds which he described as running in a haphazard way over the surface of the blastoderm, giving it an appearance which he alluded to as a 'shagreen'. Pasteels (1937 b), two years later, also thought he had found evidence, in sections, of such a 'polyinvagination' but, as we have seen, denied the reality of the Patterson type of inrolling at the margin. Jacobson (1938 a) also gives partial agreement to a similar view. In his opinion, the formation of endoderm begins by a polyinvagination of single cells, but he believes that this is confined to the posterior part of the blastoderm, instead of occurring all over it as Mehrbach and Pasteels suggested. At later stages, however, Jacobson claimed that the invagination takes a form which no one else had seen previously, namely, a massive movement of a coherent tissue from the upper layer into the lower, this movement occurring through a definite

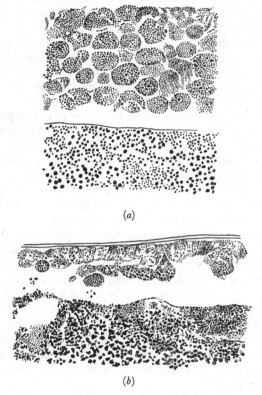

(a)

(b)

Fig. 2. Vertical sections of the blastoderm of the duck. *a* The thick blastoderm shows no sign of the separation of two distinct layers. *b* The blastoderm has expanded and the cells are becoming arranged as two layers. (After Pasteels.)

blastoporal opening in the posterior region. This process would, at least at first sight, make the bird embryo seem very similar to the reptile in its mode of endoderm formation. The novelty of the suggestion soon led other authors to make a careful search for confirmatory evidence, and Peter (1939) and Pasteels (1940, 1945) have discussed it in detail. Both are agreed that appearances such as those on which Jacobson bases his interpretation can only be produced as artifacts resulting from shrinkage during fixation; and they conclude that there is no valid evidence for anything at all resembling an open endoderm-blastopore. These further studies of Peter and Pasteels also convinced them that the same criticism must be applied to the shagreen effect of Mehrbach. As Peter (1938a, b) first pointed out, these folds are also artifacts, although Pasteels (1940, 1945) states that during late stages of endoderm formation there are certainly some cells which move between the upper and lower layers; but they are few in number, and the evidence is inconclusive whether they are moving from epiblast to hypoblast or vice versa.

Our present views as to the appearance of the hypoblast are based mainly on the memoirs of Peter (1938a, b) together with the later confirmatory studies of Pasteels (1945). Peter ascribed the formation of a lower layer to a delamination, that is, to the appearance, enlargement and fusion of clefts within the mass of cells constituting the blastoderm (Fig. 2). There is then no true invagination. The lower layer is not formed by a movement downwards of cells originating in an upper layer. It forms in situ, becoming a definite layer merely by the appearance of a space above it, separating it from the epiblast. Pasteels likens this space to the blastocoel cavity in the Amphibia, which is also a cleft appearing between the cleavage cells; if that homology is accepted, it implies, firstly, that the subgerminal cavity is not a blastocoel, as it has often been considered, and, secondly, that in the birds the blastocoel floor is directly incorporated into the embryo as the endoderm, a situation in some contrast to that found in most urodeles, but paralleled in some of the amphibians with highly yolky eggs (certain Anura, Gymnophiona). The question is further discussed on p. 48.

The process of delamination is not simultaneous over the whole germ. It begins around the margins, in the germ wall where the blastoderm lies directly on the yolk. From there it extends over the subgerminal cavity first in the posterior region. Peter describes an early stage (Fig. 3) in which the coherent layer extends only a little way forward from the posterior germ wall; anterior to it is a zone in which only isolated endoderm cells are present, while in the anterior part of the germ there is none. The question arises whether the coherent endoderm spreads into the more

anterior regions by sliding forward bodily under the epiblast, so
that there is a relative movement between the two layers. We
shall see that a forward movement of some kind can be detected
in the blastoderm at this stage. But confining ourselves to the
immediate point at issue, Peter claims (and Pasteels, 1940, 1945,

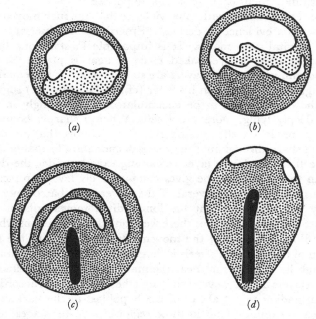

Fig. 3. The spreading of the hypoblast before and during primitive streak
formation. In the closely dotted area the hypoblast is continuous, in the
loosely dotted area it is discontinuous, and in the white areas absent. (After
Peter.)

agrees) that this does not involve any relative movement of one
layer under the other. The extension of the coherent endoderm
over the whole germ is, according to these authors, due to two
processes; first, the isolated cells in the intermediate zone lose
their original spherical shape, and become thinner and larger in
area, thus filling up the spaces between them and their neighbours;
secondly, the whole posterior region, including both layers,
participates in a forward movement which carries the coherent
hypoblast (with its accompanying epiblast above it) into more

anterior regions. This account is not entirely satisfactory, since it appears to provide no mechanism by which the endoderm-free region in the anterior can actually be reduced in area; possibly the missing part of the process is a proliferation of endoderm cells from the edges of the germ wall but it might also be a relative forward movement of the posterior hypoblast.

Peter (1939) denies with some violence that any such movement occurs. His evidence is derived primarily from observing the edges of vital-stain marks. It is impossible to stain one layer electively; when a dye-soaked piece of agar is placed on the blastoderm, usually both layers are stained more or less equally, although the figures given by Peter (cf. his Pl. 1, fig. 21) suggest that the colour tends to be accumulated more strongly in the more deeply lying, more yolky cells. When a sharply bounded mark is made on the blastoderm, Peter claims that its edge remains sharp throughout the process of endoderm formation, the dyed cells in the lower layer remaining exactly under the dyed ones in the upper; but he gives comparatively few examples to demonstrate this, and several of them were incubated only for a very few hours, so that the longitudinal movements in the epiblast had not yet begun: there were only four specimens which lived into this period of the movements which lead to the formation of the streak. Pasteels is, in general, of the same opinion, although he suggests that near the margins of the area pellucida the two layers do not move exactly together, the epiblast according to him tending to get ahead of the hypoblast in the backwards directed streaming found in those regions. At later stages, also, we shall see that Pasteels considers it necessary to distinguish between general growth changes, which affect both layers simultaneously, and characteristic morphogenetic movements of the epiblast, which would necessarily cause some relative displacement of the two layers.

The opposite point of view is represented by Spratt (1946). This author, using carbon particles as a method of marking points on the blastoderm, states that of ten blastoderms marked on the under surface, every one showed a faster forward movement of the hypoblast than of the epiblast. He relates this movement to the phenomena discovered by Waddington (1932, 1933a) in which a hypoblast, placed in an abnormal orientation, influences the direction of elongation of the streak in the overlying epiblast;

but, attractive though it might be if it could be shown that this hypoblastal inductive power is connected with the performance of certain movements in that layer, all that can safely be deduced from the experiments which have been made up to the present is that the endoderm possesses a polarity which it can transmit to the epiblast, and this polarity need not necessarily involve any movement of the hypoblast (Cf. Chapter IV).

If the forward movement occurs as Spratt claims, it would at first sight seem remarkable that it was not noticed by Pasteels in his vital-staining experiments. Spratt, indeed, supposes that Pasteels' results did show the movement, but were misinterpreted; he suggests that the forward movement described by Pasteels as occurring in both layers during the formation of the primitive streak (e.g. Pasteels, 1937 b, figs. 9, 10) is actually a movement of the hypoblast, rather than of both layers simultaneously, as Pasteels thought. This involves the implication that the dye is concentrated preferentially in the lower layer, a not impossible, but by no means a certain, eventuality. It should be possible to settle this by a special investigation which seems to be urgently called for. We shall return later to the question of these movements during streak formation (p. 23). Here it can only be said that at the present time it appears to be impossible to reconcile completely the data of Peter and Pasteels on the one hand and Spratt on the other; until that can be done, it would appear to be unsafe to rule out the possibility that there is a real shifting forwards of the endoderm relative to the epiblast, and that this contributes to the obliteration of the original endoderm-free area.

This is perhaps the place to discuss whether the delamination described by Peter is the sole source of endoderm, or whether another contribution is made to it later from the primitive streak. Hunt (1937 b) described cases in which nile blue marks were made near the side of the streak, and blue-dyed cells later found (in sections cut with a knife under the binocular dissecting microscope) to be lying in the endoderm. Peter (1939) pointed out the obvious uncertainties in such a technique, and was able to find a method of preserving the vital marks in paraffin sections. He figures isolated blue cells very similar to those described by Hunt, but he suggests that they have been coloured *in situ* by the diffusion of the dye through the depth of the blastoderm, and have become isolated from the stained cells originally above them by chance movements of the layers during the histological preparation.

The presence of such isolated stained cells might be due to the selective affinity for the dye which, according to Spratt (1946), is shown by some cells. This author has seen carbon particles, originally placed on the upper surface of the epiblast, carried down into the endoderm; but he points out that, since the particles may be transported not only within the body of a cell but also when stuck to its surface, it is possible that the endoderm cells have picked them up from the surface of an invaginated mesoderm cell. Hunt (1937a) also found that, on the chorio-allantoic membrane, grafts deprived of hypoblast could develop endodermal tissues, and argued that this indicated that these tissues normally arise from the streak; to which Peter countered quite correctly that isolated fragments are notoriously able to differentiate into more than their presumptive fate. Thus the arguments for and against a contribution of the streak to the endoderm do not seem very convincing on either side. Pasteels (1945) accepts the negative conclusions of Peter, but suggests that possibly cells move from the mesoderm into the endoderm in the region of Hensen's node at a later stage, during the retreat of the primitive streak. Again, there is no positive evidence for this, although there is no doubt (cf. Hunt, 1937b, fig. 4) that the two layers are very closely associated at the node; and Pasteels' suggestion seems to be based simply on the general grounds that the streak is, in some ways, to be considered as a blastopore, and, therefore, on comparative grounds, might be expected to yield endoderm.

The formation, development and disappearance of the streak

The first investigations of the chick blastoderm using the newer techniques of vital staining and cinema-photography were those of Wetzel and Gräper in the late 1920's and were concerned mainly with the processes occurring in the primitive streak, after at least the main mass of the hypoblast had been formed. However, in spite of the concentration of many workers on these stages, many points still remain under dispute.

Wetzel's and Gräper's researches, summarised in their papers of 1929, showed a remarkable degree of agreement in broad outline. Both stated that a movement started shortly (some 4 or

5 hours) after the beginning of incubation.[1] This movement consisted in a streaming forward along the mid-line, while the lateral parts of the area pellucida moved backwards; Gräper spoke of the formation of two vortices, comparable to one of the dance figures in a polonaise (Fig. 4). It is during this process that the primitive streak appears, and both authors agreed that it is formed out of material which originally lies rather widely spread on each side of the mid-line in the posterior sector, but which becomes drawn into an elongated streak in the median plane by the polonaise streaming. This phase, they claimed, was succeeded by one in which the main movement of tissue was from the sides

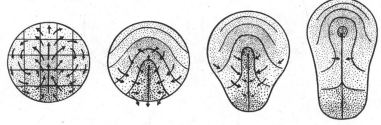

Fig. 4. Morphogenetic movements during gastrulation, according to Wetzel.

towards the streak, which had now become a definite groove. They showed that the material which reached the streak from each side then sank beneath the surface and moved out again, moving anteriorly as well as towards the side from which it had come. It lay between epiblast and hypoblast, and, in fact, constituted the mesoderm. Finally, the period of the primitive streak closed with a phase in which the tissue stream in the mid-line was towards the posterior, the streak thus decreasing in length and leaving the organs of the embryonic axis (neural plate, chorda, somites, etc.) to the anterior of it as it retreated.

The general outline of this account was accepted by Waddington (1932) as a basis for the interpretation of the results of operations on explanted blastoderms. Again, in the same general way, agreement was expressed by Pasteels (1935 b, 1936 a, 1937 b) who made a careful examination of the morphogenetic movements by a similar vital-stain technique to that used by Wetzel. Pasteels,

[1] Twiesselmann (1938) claimed that it started later, but Pasteels (1937 b) has pointed out that this was due to the cooling of his eggs.

however, added several points of major importance, as well as some details which will be considered later. In the first place he found that marks of bismarck brown dyed only the non-living vitelline membrane, and thus remained stationary to serve as points of reference. Further, he was able to distinguish two types of movement, one which affected the whole thickness of the germ simultaneously, and a second, recognised by the fact that it gives rise to peculiar striations in the colour, which concerned the epiblast alone. The former are the results of the general changes

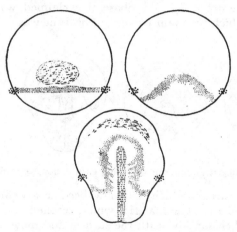

Fig. 5. Movements of a vitally stained mark placed across the blastoderm before the appearance of the streak (fine dots). The coarse dots indicate marks on the vitelline membrane which serve as reference points. (After Pasteels.)

of shape produced by localised growth, the latter are the morphogenetic movements proper. Pasteels has given a series of diagrams in which the two sets of deformations are summarised separately (Fig. 6).

There are two important points in which this Gräper-Wetzel-Pasteels scheme seems at first sight to be contradicted by more recent results. These both relate not to the transverse movements, but to the streamings along the streak: the forward movement during its formation and the backward movement during its regression. Both are derived from the work of Spratt, who followed the movements in the epiblast by tracing the position of small particles of carbon placed on blastoderms cultivated *in vitro*, a method first employed, combined with cinema-photography, in some unpublished work of Waddington and Abercrombie in 1937.

Spratt (1946) distinguished two phases in the elongation of the primitive streak: an early period, from its first appearance until it has elongated so as to reach the middle of the area pellucida, and a later period, from that time until it attains its full length.

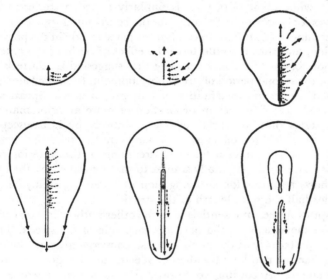

Fig. 6. Morphogenetic movements during gastrulation, according to Pasteels. Dotted arrows indicate movements which occur in the epiblast alone, full arrows movements which affect all layers. (After Pasteels.)

Spratt claims that the streak is at first about 0·5 mm. long; Jacobson (1938b) states that it can be recognised in sections when even shorter, only 0·15 mm. By the time it reaches the centre of the blastoderm (medium-broad streak, end of first period) it will have grown to a length of about 1·17 mm. according to Spratt. Now Pasteels appears to imply that this elongation is entirely due to a forward streaming, in which the material which constitutes the anterior end of the very young streak would still be at the anterior end of the medium-broad streak, while other material, originally lying to the sides, will have been added to the posterior end of the streak. Spratt claims that his carbon marks show that this is by no means the case. He finds that an important factor in the

initial elongation of the streak is the initiation of tissue condensation and invagination in front of the original anterior end, so
that the streak increases in length by incorporating new material
into itself, its original anterior end finishing in a position somewhere near the middle of its length. Spratt's measurements lead
to the conclusion that about 0·37 mm. is added anteriorly in this
way, while about 0·13 mm. is similarly added posteriorly. This
leaves only about 0·17 mm. to be contributed by a general
stretching. Spratt finds that this takes place mostly in the posterior
region, and thus it would have the effect of pushing the anterior
region forward in a way very like that suggested by Pasteels.

In the second period of streak development, from the 'medium-
broad' streak to the definitive or fully grown streak, Spratt finds
no evidence of further incorporation of more anterior material,
but states that the streak elongates entirely by intussusceptive
growth which is concentrated mainly in the posterior region,
and therefore causes a forward streaming of the anterior part.
This movement, essentially similar to that described by the other
authors, accounts for an elongation of about 0·9 mm., i.e. just
under half the total length of the streak.

Spratt agrees in essentials with the other authors in his account
of the movements of the parts on each side of the streak. These
move posteriorly at the sides, and also converge towards the mid-
line, like the two lateral halves of Wetzel and Gräper's polonaise
movement. According to Spratt, the movement is somewhat
more active than Pasteels allows (Fig. 7).

The points of difference between Pasteels' and Spratt's accounts
of the elongation of the streak are, first, a qualitative one, in that
Spratt supposes that material anterior to the very early streak
becomes incorporated into it, and, secondly, a quantitative one
in the extent of the forward streaming. As regards the former,
part of the difference is probably due to the consideration of
somewhat different stages by the two authors. The incorporation
described by Spratt takes place in the very early stages of streak
development. Pasteels' view would imply that the condensation
of material, to form a thickening which can be recognised as the
streak, begins at the extreme anterior end; Spratt's view is that
it is first noticeable somewhere about the middle. Now Pasteels
has himself pointed out that, in the somewhat later, but still
young, streak, invagination is most active in the middle and

posterior parts (1937 *b*, fig. 12 *b*). It would not be surprising, therefore, if Spratt's suggestions were correct. A similar incorporation of anterior material was described independently by Jacobson (1938 *b*) on the basis of histological study; in fact, this author was inclined to attribute the whole growth in length of the streak to such activity, which certainly gives it too great an importance. We may, however, conclude that a certain amount of anterior incorporation by the very early streak is rather more probable than not.

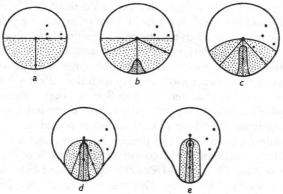

Fig. 7. Movements of carbon particles on explanted blastoderms, according to Spratt. (From Spratt.)

Spratt also differs from Wetzel and Pasteels in supposing that the forward streaming is less extensive in the early period than these authors deduce from their markings with vital dyes. Considerable evidence about the movements of the dye marks has been published, e.g. Fig. 6; see also Wetzel (1929 *a*, figs. 78–83) and Pasteels (1937 *b*, figs. 6–11). If Spratt is correct, some other explanation has to be found for these phenomena than the obvious one that they exhibit the streaming postulated by the earlier authors. Spratt advances two points. He suggests (1946, p. 287) that the dye marks are actually in the hypoblast, not in the epiblast as Pasteels thought, and that they, therefore, indicate, not a morphogenetic movement leading to streak formation, but a continuation of the forward movement of endoderm for which, we have seen above (p. 16), there is some evidence. Wetzel, however, discussed this point specifically on p. 267 of his 1929 paper, and states that the marks are certainly in the epiblast, although he admits that the hypoblast may be stained as well. In view of this, it is difficult to accept Spratt's contention. Secondly, Spratt (1947c) points out the undoubted technical limitations of vital staining in the chick, a matter to which Jacobson (1938b) had also drawn attention. Thus Spratt states that even when blastoderms (at a later age) are stained *in toto* with nile blue, there is some differential concentration of the dye. However, this remains a criticism in general terms, and does not provide a definite account of how such pictures as Pasteels' fig. 6 or fig. 10 are produced; it would not appear easy to account for them in terms of a transport of dye from cell to cell without postulating some very orderly underlying set of processes. On the other hand, it must also be remembered that it has not been fully proved that the carbon marks always move strictly with the cells to which they are supposed to be attached; thus in Spratt (1946), fig. 9 shows a mark which is not closely adherent to the blastoderm, so that it looks possible for the tissue to flow under or round it.

It appears impossible at the present time to effect a complete reconciliation between the accounts of the growth of the streak given by Wetzel, Gräper and Pasteels on the one hand, and that given by Spratt on the other. One can only conclude that either the vital dye or the carbon particles are less stable as markers than their users believe, or that the *in vitro* cultivation of Spratt's blastoderms causes abnormalities in the morphogenetic movements in spite of allowing a normal embryo to develop. It certainly seems to be the case that Spratt's blastoderms, like others explanted *in vitro* (p. 6), underwent relatively little growth in comparison with the differentiation which they performed. Now we know that in Amphibia considerable regulation of the gastrulation movements is possible, so that a harmonious neurula is produced, for instance, by embryos which have suffered a loss or addition of material. It seems very possible, therefore, that the movements in Spratt's blastoderms were not entirely normal, in spite of the fact that a normally shaped embryo was finally produced.

However, one must not exaggerate the practical importance of the differences between the two accounts. Spratt allows that 50 % of the length of the fully grown streak has been due to intussusceptive growth or streaming, while Pasteels considers the

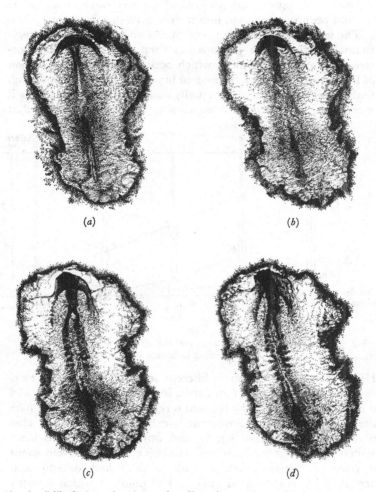

(a) (b)

(c) (d)

Fig. 8. Stills from a time-lapse cine film of an embryo developing *in vitro*, marked with six particles of carmine. Note the movement of the posterior marks towards the streak. The backward movement of the streak has been partially suppressed by the mechanical firmness of the plasma clot. (Original.)

figure nearer 70 %; the main difference depends on whether
a zone of at most 0·4 mm. in front of the very early streak has or
has not been incorporated into it during its development.

The second difference between Spratt and Pasteels as to longitudinal movements is perhaps less deep. It concerns the posteriorly directed movement which occurs during the regression
of the streak at the time the first embryonic organs are appearing.
Only Jacobson (1938b) has actually doubted the existence of such
a movement; his evidence appears to be merely the fact that

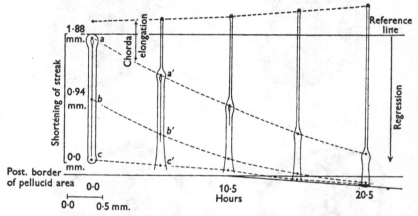

Fig. 9. The regression of the streak and the elongation of the notochord,
according to Spratt.

Hensen's node is not an indifferent mass of tissue, as Wetzel
thought, but is the site of an actual invagination (see p. 33); and
it is difficult to see how he thought it relevant to the conclusion he
drew. The backward movement is extremely clear in the cine
photos of Waddington (fig. 8), and Spratt (1947c) finds clear
evidence of it both in his carbon markings and in the development
in transection experiments of 'tails' such as had already been
described by Waddington (1932). The point in which Spratt's
account differs from the earlier ones is that he finds that the cells
of the node retreat right to the posterior end of the embryo instead
of being left behind at some point in the anterior part of the axis
(Fig. 9). This might seem to imply that the whole embryonic body
lies in front of the node, a concept which is almost certainly un-
true. We know, for instance (see p. 80), that posterior segments

isolated by cuts made behind the node can still develop neural tissue, probably not by regulative processes. The origin of this tissue must be sought laterally to the streak but posterior to the node. All that Spratt's experiments demonstrate (if they are accepted at their face value) is that there is no part of the embryonic axis lying *in the mid-line* posterior to the node. This amounts to little more than the suggestion that invagination of mesoderm through the streak continues rather later than Pasteels had thought. There does not appear to be much difficulty in accepting this, although, if it turns out that the carbon particles are not as reliable indicators of movement as Spratt thinks, it may not be necessary to do so.

During the regression of the streak, invagination movements still continue in it, even although the anterior part of the body has begun to differentiate into the definitive organs of the axis. In Pasteels' words (1940), the posterior region constitutes, for a long time, a sort of gastrula phase within an embryo of which the anterior parts are well developed.

Holmdahl, in a series of papers published between 1925 and 1939, has argued that in the chick there are two types of development. One, which may be called primary, is that which we have discussed so far, and involves a gastrulation process. According to Holmdahl, it produces only the anterior part of the embryo. The posterior part is formed by a different, 'secondary', process, of which the essential feature is supposed to be that all ectodermal and mesodermal organs develop directly out of a rapidly growing, 'indifferent' cell mass, the tail bud. The main observational evidence on which this hypothesis is based is the fact that after the formation of a certain number of anterior somites the remains of the primitive streak become very short, and slightly elevated above the surface of the blastoderm as a knob in which the cells do not seem to be clearly separated into different layers. Several authors have accepted the validity of the distinction. Peter (1934) claimed to have found the same thing in reptiles. In birds, Wetzel (1929a, 1931) considered that the greater part of the primitive streak itself consisted of such an indifferent cell mass. For Holmdahl, the boundary between primary and secondary development lay quite far anteriorly, so that the whole middle as well as posterior parts of the body arose from the tail bud by a process other than gastrulation; for Wetzel, the primitive streak was nothing but a stretched-out tail bud. Kopsch (1934), Gräper (1932, 1933) and Fröhlich (1936), on the other hand, held that the extent of secondary development was restricted to a much smaller region in the posterior. Gaertner (1949), using carbon particles to mark the cells, has recently shown that virtually all the growth in length of the embryo after the 21-somite stage is due to elongation of the tail bud.

Pasteels (1939, 1940) subjected the whole of Holmdahl's conception to very damaging criticism. He pointed out that in Amphibia, studies on the morphogenetic movements in the posterior end of the body showed clearly that the tail is formed by an invagination process extremely similar to that which gives rise to the more anterior parts (cf. Bijtel, 1931, 1936). In face of this we would need strong evidence to prove that affairs are quite different in birds; but we are offered only straightforward histological preparations, in which we are to place our faith in a negative, the impossibility of distinguishing any definite structure in the tail bud. As against this, vital-staining experiments show that the chorda of even the most posterior segments is originally located in a perfectly definite place, namely, near the node; and similarly the other organs of the posterior region originally have a definitely assignable location. It is to be presumed that they still do so in the later streak or its derivative, the tail bud; our inability to locate them exactly at that stage is not because they are an indefinite part of an indifferent mass, but simply a consequence of the small size of the remains of the streak. Gaertner (1949) also points out that the elongation takes place, not in the mass of undifferentiated tissue itself, but near the boundary between this and the differentiated region which joins the tail bud anteriorly.

Moreover, Pasteels was able to show (1942b, 1943) that the usual notion that the tail bud is the site of particularly rapid proliferation and growth is an erroneous impression, derived probably from the fact that it is formed of small cells similar rather to those of the primitive streak than to others in the differentiated parts of the body. Direct counts of mitoses established no noticeable preponderance of cleavages in the tail bud region over that found in most other organs. Gaertner (1949) confirms this. The idea of a 'secondary' type of development thus has little to be said for it; the posterior parts are formed by processes essentially similar to those which produce the anterior regions, and which have been discussed above.

It should be pointed out that Pasteels seems to be guilty of some inconsistency in this connection, since neither in his maps of gastrulation movements nor in those of the presumptive areas does he give any indication that mesoderm is still remaining on the surface and in process of being invaginated in the posterior region after the attainment of the definitive primitive streak stage. In the figures of movements, the invagination is difficult to distinguish from the elongation and condensation which he shows as characterising the whole posterior region; but in the maps of presumptive areas, a more definite indication of an area of presumptive mesoderm should be added to the diagram he has provided (see Fig. 10).

The mechanism of gastrulation movements

We have little or no positive knowledge as to the mechanisms which bring about the morphogenetic movements in the chick, but there are some negative points of interest. It has frequently been considered that such movements may be due to localised proliferation. Wetzel (1929a), for instance, held that the primitive streak is the site of such a process. However, Pasteels (1940, 1942a) could find no statistically significant differences in mitotic rate within the blastoderm, and pointed to the lack of tests for significance of the differences quoted by authors who have claimed to find them in other forms. Derrick (1937) also found no consistent difference in favour of the primitive streak region. We may conclude that differential proliferation is not one of the factors which set in motion the morphogenetic movements.

This is also true, as Pasteels shows, in Amphibia. In this group an attempt to propound a physico-chemical theory of the movements has been made by Holtfreter (1943, 1944). It is perhaps worth pointing out that although most of the mechanisms which he envisages could be supposed to act in cases where a surface is deformed, as the spherical surface of the amphibian egg is bent to form the groove of the archenteron, it is much more difficult to see how they could bring about the movements of the chick, many of which (such as the forward streaming) appear to occur mainly in one plane. This raises the suspicion that the chick movements may not, after all, be so uniplanar as they seem. It

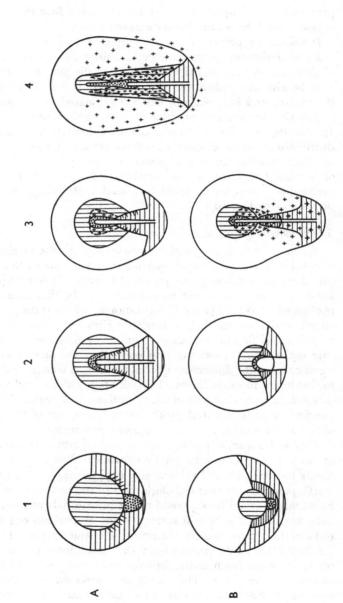

31

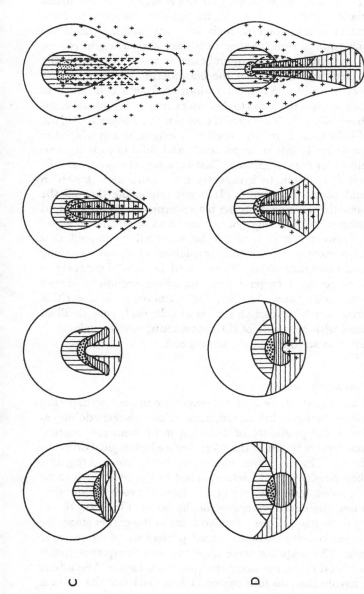

Fig. 10. Maps of the presumptive areas. Row A after Gräper (1929); row B after Waddington (1932); row C after Pasteels (1937); row D are maps suggested in the light of more recent results. Column 1, pre-streak stage; column 2, early streak; column 3, fully grown streak; column 4, head process. White represents presumptive epidermis; vertical lines, neural tissue; dots, chorda; fine dots, invaginated chorda; close horizontal lines, axial mesoderm; spaced horizontal lines, non-axial mesoderm; crosses, invaginated mesoderm.

might be the case, for instance, that the elongation of the streak is essentially connected with the invagination of cells to form mesoderm. The first stage in such an invagination is probably a tendency, due to some change in surface properties, to reduce the area in which the cell membrane is in contact with the external medium surrounding the blastoderm. We may picture the cells of the mid-line being transformed from a cylindrical to a bottle shape, with the necks to the outer surface. If such a change of cell shape is not to leave spaces between the narrow necks, the group of cells must rearrange itself in a long strand in which the necks can interdigitate so as to reach and fully occupy a comparatively small area of surface. That is to say, the group of cells would not only begin to invaginate but would also stretch in length and contract in width. It is not impossible that all the morphogenetic movements in the blastoderm will eventually find their explanation in changes of cell shape in some such manner as this. However, it cannot yet be accepted as proved that gastrulation movements are to be explained by processes taking place on the cellular level. It may well be that the causative agents operate on a larger scale, and affect regions of tissues rather than cell shape. In fact, the phenomena in the chick blastoderm, which consists of a mass of cells each very small in comparison with the scale of the movements, would make such an explanation seem the more natural one.

Maps of presumptive areas

From the point of view of the experimentalist, perhaps the greatest importance of the investigation of morphogenetic movements lies in the possibility of deducing from them the location of the presumptive tissues in the stages before histological differentiation begins. In the chick, vital-stain marks are too fugitive, and carbon particles too loosely attached to the cells, for it to be possible to follow a single group of cells from the early blastoderm through into the definitive organs of the body. One can, therefore, only draw maps of presumptive areas in the earlier stages by a more or less convincing series of back projections of the relevant movements. The maps for these stages are in consequence much less certain and much less accurate than those for the Amphibia; one can hardly hope to find the exact boundaries of the various areas, and must be content with some knowledge of their general size and arrangement. Both Wetzel and Gräper summarised their results in the form of maps, which have been redrawn in Fig. 10. Those of Wetzel indicate only a region of presumptive neural tissue which was considered always to lie in the surface, and another area (indicated by dots) which would give rise to the streak. As has previously been pointed out (p. 26), Wetzel considered the streak to be made up of 'indifferent' material which proliferated to give both mesoderm and neural tissue. Gräper (1929a) and Waddington (1932) early recognised that the streak is the site of a real invagination rather than a mere mass of indifferent tissue from which mesoderm and neural plate are given off and we have seen that Pasteels (1943) later demonstrated that no such proliferation occurs in it. In Gräper's diagrams, therefore, the presumptive mesoderm and neural tissue are separated. But his diagrams had other imperfections; the earliest marked the endoderm in a location deduced from Patterson's erroneous ideas as to its formation; furthermore Gräper held, on comparative anatomical grounds, that the streak, being a blastopore, must have an endodermal core; and, finally, his method of cinema-photography was not used in a quantitative way, and his diagrams made less pretensions than Wetzel's to being spatially accurate.

At a slightly later date, Waddington (1932) published some maps from which the most obvious faults of Wetzel's and Gräper's had been eliminated, and which also took account of results derived from defect and isolation experiments. They differ from Gräper's mainly in the omission of the endodermal core of the streak, and in drawing the presumptive axial tissues further posteriorly in the early stages.

As a result of his experiments, Pasteels (1937b) put forward another series of maps. In the middle and late primitive streak stage, these resemble Waddington's maps fairly closely. The only important new point concerns the lateral mesoderm. Waddington had accepted Wetzel's view that all the mesoderm lateral to the somites was invaginated in the posterior part of the streak and moved forward from there after invagination (this invaginated lateral mesoderm was omitted in the original drawing, since attention was concentrated on the axial organs). Pasteels, however, believes that some lateral mesoderm is invaginated through the anterior of the streak in early stages. This would mean that in the young streak stage, a U-shaped region of the area marked as axial mesoderm in Waddington's map should really be lateral mesoderm.

It has been pointed out, both by Jacobson (1938 b) and Spratt (1947 b), that the presumptive neural plate material can be recognised in stained preparations of the late primitive streak blastoderm, since it tends to stain somewhat more darkly than the rest of the area pellucida. This is partly a consequence of the fact that it is composed of a thicker and more compact epithelium (which is presumably an indication of the beginning of an inductive action by the underlying mesoderm) and partly, in Jacobson's view, a result of its higher content of lipoids.

Waddington's and Pasteels' maps of the pre-streak stage appear at first sight to differ more profoundly, but a major part of this difference is simply due to the authors having chosen to represent differently the region of the neural plate. In actual fact the shape of this area is not known with any accuracy; both maps are highly diagrammatic and only attempt to show the general location of the presumptive area rather than its actual boundaries. Pasteels' map, however, gives a much smaller area to presumptive mesoderm than does that of Waddington, although even the latter gave less than Gräper had done. Pasteels, in fact, would seem in his first diagram to have allowed too little space to the lateral mesoderm for this to develop easily into his second diagram; in particular, it seems unlikely that the arms of the presumptive axial mesoderm actually reach as near the margin of the area pellucida. On the other hand, it is probable that Waddington drew the presumptive chorda and neural plate too far to the posterior.

The arrangement of the presumptive areas is, as has been said above, deduced from our knowledge of the morphogenetic movements, and it is necessary to consider what alterations are called for to meet the points made by Spratt (1942, 1946, 1947 a, b, c). His data on the regression of the streak make no great difference to the boundaries of the main areas, since they can be interpreted in terms of the cell movements within each area during the process of elongation of the embryonic axis. If the node, as he suggests, contains material for the most posterior segments of the body, we must suppose that the intersegmental boundaries, instead of running forward obliquely for a short distance towards the streak, as Pasteels' diagram indicates, actually extend right forwards into the node, as Wetzel (1929, fig. 119) seems to have thought. It is probably easier to accept this if we concede that the invagination continues longer in the anterior part of the streak than Pasteels allows. This was the view of Gräper and Waddington, though it is not very clearly indicated on their diagrams except by the continued presence of chorda on the surface, and it is certainly the view of Spratt (1947c, p. 79). In fact, Pasteels (1937, fig. 12d), shows invagination still proceeding at a stage when his maps indicate that all the mesoderm should already be below the surface; it is suggested that the former alternative is the correct one At the same time, one must conclude that invagination takes place very slowly over the anterior lip of the node. This has already been pointed out by Pasteels, and there is little difference in this respect between his views and those of Spratt.

The positions, on this hypothesis, of the mesodermal and neural parts of some typical segments of the axis during invagination are shown in Fig. 11.

If Spratt is correct in minimising the importance of forward movements during the early development of the streak, considerable changes would have to be made in the map given by Pasteels for the earliest stage. These would, in some ways, be in the direction of bringing Pasteels' map more into line with those of Gräper and Waddington. That is to say, the presumptive axial organs would not be concentrated so near the posterior, but placed nearer the centre of the area pellucida. On the other hand, if the forward streaming in the mid-line is less than Pasteels thought, the original area occupied by the neural tissue could not be the more or less round region figured by these authors, since by the stage of the fully grown streak it must have been converted into a horseshoe-like area. Apart from the forward movement in the mid-line, another factor could assist in this, namely, the backward movement along the sides. Spratt claims that nearly the whole posterior half of the unincubated blastoderm will eventually be invaginated as mesoderm and, if this is so, the neural area might be a band stretching from right to left across the germ just anterior to the centre. It seems probable that one should accept the evidence that a greater extent of the material becomes invaginated, since it is based on the positive observation of movement rather than a negative failure to observe shiftings; moreover, there do not seem to be any experiments published by Wetzel or Pasteels which speak strongly against it. As we have seen, the evidence with regard to the extent of the forward movement is conflicting, and the best one can do at the present stage

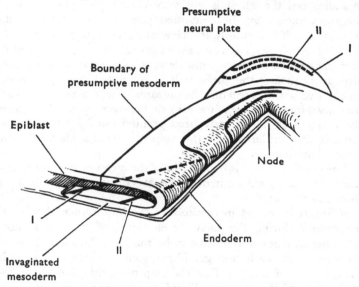

Fig. 11. The blastoderm is imagined as cut along the streak from the posterior towards the node, and the cut edges dragged apart to form an open V. Each of the heavy lines, I and II, indicates the position of a future transverse section through one side of the mesodermal part of the axis, I being the more anterior. The heavy dotted lines indicate the corresponding neural tissue of the same transverse sections.

is to draw a map on the basis of a compromise, placing the areas further anteriorly than did Pasteels, but not so much as suggested by Spratt or allowed by Rudnick (1948). Still less can one go the whole way with Jacobson (1938b), who denies both the initial forward and the later backward movement along the streak, and therefore places the presumptive areas in the unincubated blastoderm nearly at their final location. A similar conclusion was reached by Kopsch (1926, 1934) on the basis of marks made with an electrolytic needle. However, this instrument produced injuries which probably had considerable effects on the course of development and certainly made the interpretation of the results so uncertain that little weight can be given to them. We have rather little information about the details of the movement of the mesoderm after it has been invaginated through the streak. There can be no doubt that the lateral plate mesoderm moves fairly rapidly towards the sides, but it is of more importance to discover exactly what happens to the axial mesoderm. The interpretation of some of the isolation experiments to be described later depends to some extent on the magnitude of the transverse width which one can suppose these invaginated axial tissues to occupy (p. 86). In Pasteels' diagrams it is shown as forming a fairly narrow band, while in Waddington's it occupies a relatively greater proportion of the width of the area pellucida. Rudnick (1944) has given a map in which the invaginated tissues are marked as remaining almost stationary in the streak. This seems almost certainly an error. The two sketches in row D of Fig. 10 are admittedly diagrammatic, but the experimental evidence would support a wider rather than a narrower extension, since tissues such as heart, cartilage, muscle, etc., can be produced by isolates taken some distance laterally to the streak. There is little doubt, of course, that the presumptive heart tissue must move some considerable distance laterally to each side of the streak before being brought back towards the mid-line when the two rudiments become drawn together to form the heart itself. Again, the axial mesoderm, after invagination, takes part in the great posterior elongation of the body, and in this the increase in length must in the main be compensated by a decrease in width, as we know it to be in the case of the neural tissue; and from this one may deduce that the invaginated mesoderm originally has a greater width than it will have in the definitive body.

Invagination through the streak is not the sole source of the mesodermal tissues of the entire embryo. A further contribution is derived by the formation of mesenchyme from the neural crest at the time of closure of the neural tube. This has not been followed in detail by vital-staining, but normal histological observation, and also experimental evidence (p. 184) leaves no doubt that 'mesectoderm', which later forms ganglionic tissue and pigment cells, is produced in this way in the chick as it is in Amphibia. It is also sometimes suggested that mesoderm may be formed by a process akin to delamination in the region of the area opaca, but there is little evidence of this.

In Fig. 10, row D, an attempt has been made to modify the older maps of Waddington and Pasteels in the light of the considerations discussed above.

A few words are necessary about the maps of stages later than that of the fully grown streak. These can, in general, be deduced with sufficient accuracy from what has been said about the nature of the morphogenetic movements. A point which seems not always to be remembered is that the material of the chorda actually moves a small distance forward under the epiblast anterior to the node (see Pasteels, 1937b, fig. 16; Spratt, 1947, fig. 3). Thus the tip of the head-process lies under ectoderm which has never been in actual contact with the primitive pit, but has always lain anterior to it. The point seems to have been overlooked by Clarke (1936) and by Spratt himself (1940).

In still later stages, the sinus rhomboidalis retains many of the characteristics of the streak. A map of presumptive regions in it has been furnished by Wolff (1936), based largely on the results of defects made by irradiation with a narrow beam of X-rays. The situation in still later stages, when the tail bud has appeared, is considered on p. 174.

CHAPTER III

COMPARISON WITH OTHER VERTEBRATES

A T many points in other chapters mention is made of phenomena in other groups of vertebrates, particularly in amphibians, which seem comparable to those which have been discovered in the avian embryo. There is no doubt that the development of the amphibians has been more fully explored, both from the descriptive and the epigenetic points of view, than that of the birds, although the other vertebrate groups are less well understood. One can, however, make at least a beginning of a comparative account of epigenetics within the phylum as a whole, and comparisons between the two better studied types can be instituted in some detail.

Recapitulation

In the early period with which we are mainly concerned, little use can be made of the famous 'law of recapitulation'. As many people have pointed out, and de Beer (1940) in particular has emphasised, this so-called law really indicates no more than that the processes of development of an ancestral form are frequently found to be repeated, in altered but still recognisable form, in the life history of a descendant. It is not a principle which is invariably applicable; and it is certainly not a principle of cause and effect. In point of fact, we find that the fertilised ovum presents very great differences in the various vertebrate groups, particularly in the quantity of relatively inert yolk which it contains; and, clearly correlated with this difference, amongst others, the earliest stages of development are greatly dissimilar. They approach one another more closely later on, during the post-gastrulation stages. The conformation of the embryonic body in a tail bud amphibian is more like that in a chick with a similar number of somites than is a newt gastrula to a primitive streak blastoderm.

Such early dissimilarity and later convergence present considerable difficulty to evolutionary theory. If we imagine that a certain stock has become permeated with a genetic change which alters some of its early developmental processes, it is not immediately obvious in what way it is brought about that the later developmental stages are relatively little changed. In fact, such a phenomenon would seem inexplicable unless we conceive of epigenetics as essentially involving equilibrium phenomena. If it may be imagined that, in some way that we do not yet understand, the pattern of the vertebrate embryo of the somite stage is the expression of an equilibrium between interacting agencies, then it becomes comprehensible that a similar pattern may be attained even after some of the earlier stages have been altered. The hypothesis that the patterns which we find to be repeated in the development of different groups are equilibrated ones, besides being called for by the considerations just mentioned, suggests other consequences which can be experimentally tested. We should, for example, expect to find considerable evidence of a capacity in such patterns to act as teleogenetic ends in Dalcq's sense, that is, to tend to be restored by regulation after defects or other disturbances had been made to earlier stages. In the section on individuation (Chapter VI) we shall see that, in fact, in the chick there is a considerable tendency for such teleogenesis, although there are many experimental situations in which normality is not fully restored. But the period in which regulation is possible is exactly that which we should expect, namely, that in which the fundamental embryonic pattern is first being attained. The regulative capacity thereafter falls to a minimum; when it rises again, it is the pattern of individual organs which is the equilibrium to which the disturbed system tends, and no longer the pattern of the embryo as a whole.

The same hypothesis might also suggest that we should find that the majority of genetic changes produce their effects without disrupting the patterns which we have considered to be equilibrated. In practice we know too few genetic changes affecting early stages to be able to form an opinion on this point. But it should be noticed that the hypothesis of epigenetic equilibration actually does no more than suggest that there should be some genes which alter development but leave the fundamental patterns intact. It is in no way controverted by the existence of mutations which disrupt these patterns as, for instance, one sees in the short-tailed mutants studied by Schönheimer-Glücksohn in the mice. For there is another point to be remembered. If one considers an evolving epigenetic system as a variable affair, continually modified by the influence of mutation and selection, then each basic equilibrated pattern will provide a point of relative stability, with respect to which the immediately subsequent set of processes can be organised. Changes in the pattern will not only tend to lack the quality of equilibration, and thus to be subject to a considerable amount of variation, but they are likely to be of particularly great consequence to the whole of later development, which will have been organised on the basis of a pattern which has remained stable through perhaps a very large series of evolutionary advances. One must, therefore, expect that genetic changes which affect these basic patterns will only in an infinitesimal proportion of cases be suitable for incorporation into the genetic make-up of the species. They might, of course, be of some evolutionary importance if they affect only a restricted and comparatively unimportant region of the body, such as the tail-less factors just mentioned. But the argument renders it easy to understand why it is that in all the divergent groups of vertebrates, we find none which develops three rows of somites, or only one row with the suppression of the notochord, or any of the many other general changes of embryonic pattern which would at first sight seem easily attainable.

One can point to some actual examples in which the dependence of later processes on the integrity of an earlier pattern is experimentally demonstrable. It is sometimes argued that the chorda is retained in the development of backboned animals because it is required to induce the neural system. But this is not very convincing. In the first place, the neural system is induced, not by the chorda as a separate organ, but by the axial mesoderm as a whole. In the Amphibia, at the time of the inductive action, the mesoderm is still in the form of a continuous sheet of archenteron roof. Its separation into three main strands, of chorda in the middle with somite on each side, is not essentially concerned with the induction of the neural system as a whole. Its influence on that system is confined to the production of the characteristic shape of the neural tube, with its thick walls and thin floor. This shape, however, becomes considerably modified in later development, and there would seem to be no compelling reason why the 'thick-walled, thin-floored' stage should ever be undergone in the way it is, unless we admit the consideration that the pattern of the early embryonic body is one of the equilibrated reference points around which the evolving epigenetic system has been organised. But we cannot use this example as an argument to demonstrate the importance of the equilibrated patterns, since the only known effect of the pattern is one which we cannot show to be essential.

A more convincing example can be found in the epigenetics of the nephroi in the chick. The pronephric rudiments are as functionless as excretory organs as the chorda is as an organ of support. But, as we shall see (p. 177) the pronephric duct plays an important, if not an essential, part in the induction of the mesonephros, the function of which is necessary to the metabolic life of the embryo. We seem to have here a real case of the morphogenetic function of a vestigial organ (Waddington, 1938a).

Comparison of presumptive maps

It seems unnecessary to discuss at length the differences between the cleavage patterns in the birds and those in other vertebrates, since in the former the cleavages are not very well known nor do they appear to play any important epigenetic role. The stage which it is most essential to bring into a general scheme of comparisons is that of the blastula and gastrula, which have been of such great importance in the comparative embryology of earlier times. As Pasteels (1940) has emphasised, it is useless to attempt to understand gastrulation by establishing a series of homologies between structural forms. The phenomena in which one is interested are essentially dynamic ones, and forms which have a certain topological similarity may be dynamically of quite different significance. Thus the two-layered avian blastoderm at a time just before the appearance of the primitive streak cannot in any important sense be considered equivalent to Haeckel's gastraea; the space between its two layers is not equivalent to an archenteron, as normally defined; and, indeed, if we adopt structural definitions we cannot even find anything representing such a fundamental entity as the blastopore, which we should have to consider as a pore opening into a closed gut-cavity. Instead of seeking for the homologues of certain static shapes, we have to institute comparisons between the elements of a dynamic and changing pattern. These can best be discussed in terms of the maps of presumptive areas and of the movements by which these areas are brought into their final positions.

Although the amphibians still remain the group in which the location of the presumptive areas before gastrulation has been most thoroughly studied, we now have a sufficient body of data concerning the other classes of vertebrates on which to base a general picture of conditions in the phylum as a whole. In fish, we have results on teleosts, from Pasteels on *Salmo*, and Oppenheimer on *Fundulus*; on Selachia, by Vandebroek on *Scyllium*; and on cyclostomes by Weissenberg; while in the reptiles Pasteels has published important studies made with vitally stained marks, and Peter has made extremely careful investigations using older methods. Finally, in the prochordates Vandebroek has added further precision to the older data of Conklin and others. General reviews of these types have been given by Pasteels (1940) and in less detail by Dalcq (1938, 1941).

In the prochordates, the cyclostomes and the urodele amphibians, the egg is not charged with very great quantities of yolk, the cleavage being total and a hollow, more or less spherical, blastula being formed. The similarity in the basic patterns of the presumptive maps is obvious at a glance. The blastopore lies within the endodermal area, but usually near the boundary between it and the mesoderm. Proceeding along the dorsal surface away from the blastopore, one passes successively into the zones of mesoderm, neural plate and finally epidermis. Within the mesodermal zone, the axial mesoderm (chorda and somites) is located nearest the dorsal plane, and is the first part of this tissue to become involved in the invagination. The mesoderm may extend on the ventral side right round the surface of the blastula, in the form of presumptive lateral plate material; in other cases, as in cyclostomes, presumptive mesoderm may be lacking in this region and presumptive ectoderm and endoderm come into contact with one another.

The avian egg is, of course, very markedly telolecithal, and before attempting to make comparisons between its presumptive map and this fundamental pattern of the lower vertebrates, it will be as well to consider the situation in the telolecithal anamniotes. Pasteels (1949) and Nieuwkoop & Florschütz (1950) have recently described the process of gastrulation in the moderately yolky eggs of the anuran *Xenopus*, but unfortunately, we do not have accurate information on the moderately telolecithal forms such as the more yolky amphibians (Gymnophiona, etc.) or fish (e.g. such

forms as Amia). We must pass directly to the extremely telolecithal teleosts. The picture they present is, however, quite easily
interpreted. The map of a teleost blastoderm is exactly that which
would be expected if an amphibian blastula were opened up at
the position of the blastopore, and the mass of yolk inserted there.
That is to say, the teleost blastopore is situated at the posterior

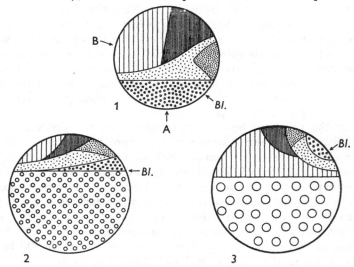

Fig. 12. Maps of the presumptive areas in blastula stage: 1. amphibians,
2. teleosts, 3. birds and reptiles. Wide vertical hatching, epidermis; close
vertical hatching, neural plate; spaced dots, mesoderm; close dots, chorda;
small circles, endoderm; larger circles, yolk; *Bl.* position of blastopore. In
fig. 1, *A* indicates the place into which the yolk-mass would have to be inserted
to produce the condition found in teleosts, while *B* indicates the point of
insertion required to produce the condition in sauropsids.

margin of the blastodisc; immediately in front of it, and stretching
round to the sides, is a crescent of presumptive endoderm, inside
which is a ring of presumptive mesoderm, followed by the ectoderm
in the centre, with the neural ectoderm placed dorsally near the
blastopore.

In the sauropsids and probably in the amniotes as a whole, the
arrangements of areas is quite different. The most striking alteration is that the presumptive ectoderm, instead of being placed
centrally surrounded by a ring of other tissues, is here peripheral,
and forms the actual edge of the blastoderm. We might say that

in the reptiles and birds, the mass of yolk has been inserted into a position which corresponds, not with the blastopore of the amphibians, but with an opening imagined to be made in the presumptive epidermis. This is, in essentials, the relation which Pasteels (1940) believes to hold between the sauropsidan and the anamniote presumptive maps. But the theory is not altogether satisfactory. The differences between the amniote and anamniote patterns are more profound than it suggests, since there are two other important points which need consideration. These are the situation of the presumptive endoderm and the relation between the centre of invagination and the various regions of the presumptive mesoderm.

As regards the former, our knowledge of endoderm formation in sauropsids is still, as we have seen (p. 12), by no means satisfactory. It appears, however, that in the birds endoderm formation takes place by delamination *in situ*, so that the presumptive endoderm never appears on the surface of the blastoderm. There is no system of opening out a urodele blastula, and inserting the yolk, which could yield an appropriate map for this situation, since any such deformation will leave the presumptive endoderm on the surface. We have to conclude that in the birds the mass of yolk does not come in contact with a cavity which corresponds to the amphibian blastocoel, although in the teleosts and selachians, that blastocoel is represented by the subgerminal space. In birds, as Pasteels has pointed out (1945), the so-called subgerminal space is really merely a mass of liquefied yolk which is normally lost during histological preparation; but it is in no sense a cavity and does not represent the blastocoel. The latter can be found only in the thin split between the epi- and hypoblast of the two-layered germ. It is into this space that the invagination of mesoderm takes place through the primitive streak, just as mesoderm is invaginated as the archenteron roof into the blastocoel of an amphibian. But we have then in the chick no process corresponding to the invagination of the floor of the archenteron in such a form as *Triton*. It seems probable from such data as are available that in some of the more strongly telolecithal amphibians the floor of the blastocoel may be directly incorporated into the gut endoderm without going through a real invagination, and if that is the case, the situation among them would be more nearly parallel to that in birds.

It may be admitted that in choosing birds as our representatives of the sauropsids, we have taken an extreme type. In certain reptiles, such as the lizard and snake studied by Peter (1938c), the formation of the endoderm seems very similar to that in birds, but in the Algerian turtle (*Clemmys leprosa* Schw.) Pasteels (1937) showed that some of the endoderm is originally on the surface of the blastoderm, and is invaginated through a true blastopore. In such a form the presumptive map of the blastoderm surface is indeed very like that which would be obtained by inserting a large mass of yolk in the presumptive epidermal region of the amphibian blastocoel, in the manner suggested by Pasteels; but even in *Clemmys* a considerable portion of the endoderm is derived by delamination from the deeper layers, where its presence would not be expected on such a hypothesis.

The other fact about sauropsidan gastrulation which does not seem easy to account for on the Pasteels scheme is the fact that the first mesoderm to be invaginated is not axial but lateral mesoderm, which moves to the outer edges of the lateral plate. This is as though the ventral lip of the amphibian blastoderm were formed before the dorsal lip.

It is possible to suggest another derivation, alternative to the Pasteels one, by which the sauropsidan presumptive map could be related to that of an anamniote. Since this scheme attempts to take account of the two factors mentioned above, it involves a more complicated set of distortions than those suggested by Pasteels. The essential feature of it is that, in order to account for the position of the main mass of endoderm which is formed by delamination, it imagines a situation in which the yolk granules are removed from all the cells of the amphibian blastula and accumulated as a non-cellular mass towards the vegetative pole, while the now yolk-free cells of the blastula are pressed together in the vertical direction, so that the blastocoel cavity is obliterated, and the roof squeezed down into contact with the floor. In this way the spherical amphibian blastula would be reduced to a flat disc of moderate thickness, the edge of which would represent a more or less latitudinal line traced somewhere below the equator of the original sphere. The exact position of this line, which we may call the 'primitive edge' of the blastoderm, will need discussion. But before engaging in that, it must be pointed out that a further distortion must be considered necessary. Any such flattening as we have contemplated will leave presumptive mesoderm (or possibly endoderm) bounding the edge of the blastoderm on the dorsal side, as it does in telolecithal anamniotes. In sauropsids, this edge is occupied by ectoderm. In the present scheme, the origin of this represents a second step in the derivation. To supplement the flattening of the germ, we have to postulate an

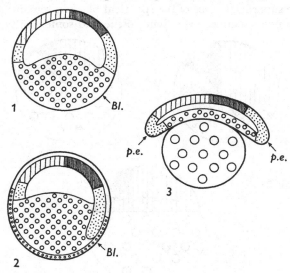

Fig. 13. 1. A vertical section through a blastula of a urodele, showing the presumptive areas with the same conventions as in Fig. 12. 2. The condition in very yolky amphibians, such as *Xenopus* (unshaded, surface layer of ectoderm; small circles, surface layer of endoderm). 3. Section through a hypothetical blastula in which the yolk-mass (large circles) has been segregated from the endoderm (small circles) and the blastocoel roof pressed down towards the floor, largely obliterating the blastocoel cavity.

extensive, somewhat epibolic expansion of the area of presumptive epidermis, which not only carries it radially beyond the primitive edge but also causes it to sweep round from the two anterior-lateral regions towards the posterior, so as entirely to enclose the original dorsal region where the primitive edge was made up of presumptive mesoderm.

This imagined expansion of the uppermost, ectodermal, layer of our hypothetical blastula is not so unparalleled as it might seem at first sight. Something of a rather similar kind seems to occur in the moderately yolky Anura, such as *Xenopus* (Pasteels, 1949; Nieuwkoop & Florschütz, 1950). In these, the presumptive fate of a region of the blastula is not the same all the way through the thickness of the tissue, so that the map must be considered in the full three dimensions, and cannot be reduced to terms of merely the surface of a sphere. The first peculiarity of this form is that the superficial layer of the spherical blastula remains always as a boundary between the living tissue and some empty space,

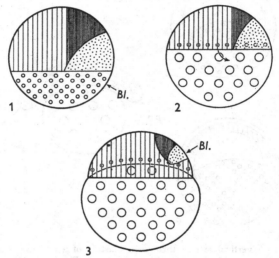

Fig. 14. The derivation of the sauropsidan presumptive map. 1. A blastula is shown, in lateral view, in which the mesoderm does not extend to the ventral side. 2. The yolk-mass is imagined as segregated to the lower pole (large circles) leaving the endoderm (small circles) hidden beneath the other tissues. 3. The presumptive epidermis is imagined as having expanded in area in the direction shown by the arrow in 2.

either external or internal to the animal. Thus part of this layer becomes invaginated, but forms then only the lining of the internal gut-cavity and never becomes buried between other tissues, as does the mesoderm of a urodele. The surface of the blastula is thus occupied only by presumptive ectoderm and presumptive endoderm, the presumptive mesoderm having no place on it. All the presumptive mesoderm lies deeper. It is probable that some of it is covered by the thin outer layer of presumptive ectoderm, which can be regarded as having expanded beyond the limits oi its 'normal' position, in the way suggested above; but the boundaries of the presumptive mesoderm and ectoderm have not been determined accurately enough for one to be certain that the latter overlies the former. However, the relations between presumptive mesoderm and endoderm admit of less doubt. It appears to be incontestable that the mesoderm extends downwards over the endoderm and directly covers it. Here there is no question but that, if at about the level of the equator one passes inwards from the surface in a radial direction, one would (after passing the thin superficial layer whose exact fate is doubtful) be passing from cells with a more 'animal' fate to ones with a 'vegetative' fate. It is to an exaggeration of the same type of distortion, by which this condition has been derived from that of the urodeles, that one may look to find an explanation for the expansion of the sauropsidan presumptive ectoderm.

We may now return to a discussion of the precise location of the 'primitive edge', i.e. the latitudinal line along which the vegetative part of the amphibian blastula is folded up to come into contact with the animal part. In the dorsal region, this line of folding must, in birds, lie above the limit of the presumptive endoderm, all of which is placed below the surface. In reptiles such as *Clemmys*, on the other hand, the edge must lie somewhat lower, so that some endoderm remains exposed.

From its position on the dorsal side, the primitive edge must run across the blastula to the ventral side, following more or less the boundary between presumptive endoderm and mesoderm. The exact position in which it should be placed depends on whether we allow that in the sauropsids any of the lateral mesoderm is also formed by delamination, for instance in the area opaca of birds, or whether it is all derived by invagination through the streak. The question is not entirely decided, perhaps, for normal development, but Olivo (1928) obtained differentiation of heart muscle from *in vitro* explants of the margins of unincubated blastoderms, which, unless it is due to regulation within the isolate, would suggest that some mesoderm can arise directly *in loco* in the most lateral regions. If this is so, the primitive edge must lie somewhat above the boundary of the presumptive endoderm.

In a form such as the bird, the dorsal part of the primitive edge must lie near the lower boundary of the presumptive mesoderm,

that is to say, well above the dorsal lip of the blastopore, while
further ventrally it must lie much nearer the position of the lateral
lips. It is in this relation that we may, perhaps, find an explanation
for the fact that the first mesoderm to be invaginated through the
primitive streak is lateral plate mesoderm, and that it is only later
that the axial mesoderm comes to be invaginated. This suggestion
would tend to reinforce the argument of Spratt and Jacobson that
the early streak grows to some extent by the incorporation of new
material at its anterior end (p. 21).

CHAPTER IV

EPIGENETIC PROCESSES ASSOCIATED WITH ENDODERM FORMATION

Changes in the relations between epiblast and hypoblast

The most informative investigations on the epigenetic reactions at the stage of endoderm formation are, as might be expected, those in which the relations between the hypoblast and epiblast have been altered in a definite and specific way. Unfortunately, few of these experiments have as yet been performed (Waddington, 1930, 1932, 1933c).

One difficulty in operations on the pre-streak blastoderm is that of orientation. The earliest appearance of the streak, of course, gives a firm point of reference, since this structure develops always in the posterior region. In earlier stages we have seen that the endoderm first becomes a coherent layer in the same region. This is not always easy to recognise under the binocular dissecting microscope, although in favourable cases a high degree of certainty in orientation is possible (discussion and figures in Butler, 1935, and Spratt, 1942). It is perhaps somewhat easier in the duck than in the chick. Some authors have, perforce, relied on von Baer's rule, that the embryonic axis usually lies at right angles to the long axis of the egg with the head pointing away when the blunt end is towards the left; but Butler (1935) finds that the rule is exactly fulfilled in only 50 % of the cases, while in 33 % the embryonic axis was rotated 45° or less to the right, and in about 11 % a similar amount to the left; the deviation in the remaining 6 % was even greater, in 1 % the axis being actually reversed. However, there is probably some variation in this respect between different groups of hens, since Kopsch (1926b) found only 33 % in agreement with the rule, and nearly 6 % in disagreement and other authors have given even higher percentages for the latter class (Bartelmez (1918) up to 33 %).

In Waddington's experiments both duck and chick embryos were used. Orientation was ascertained either by operating on very early primitive streak stages or by the recognition of a coherent layer of endoderm in the posterior region. In the first series of experiments (1930, 1932), the hypoblast and epiblast

were carefully separated over the whole area of the blastoderm
(or, in slightly later stages, over the area pellucida only) and the
epiblast then replaced on top of the hypoblast after rotation
through 90° about a vertical axis, and the new combination
cultivated by the watch-glass method *in vitro*. It is convenient to
describe the orientations of the various structures and organs in
these experiments in terms of a clock face. We will speak of an
embryonic axis (or part of it) 'pointing' in the direction given as
one proceeds from the posterior towards the anterior end of it.
The direction '12 o'clock' will be defined as that towards which
the presumptive axis of the epiblast points; then, for instance, if
the epiblast was originally arranged so that its posterior end was
nearest the observer and the anterior end away from him, an
embryo which developed with its posterior to the left and its
anterior to the right would be said to 'point to 3 o'clock'.

Several of the embryos in which epiblast and hypoblast had
been subjected to a relative rotation of 90° developed as normal
embryos (i.e. as normal as can be expected in culture) which
pointed in the epiblastal direction, i.e. to 12 o'clock. They were
furnished with a normal fore-gut, and one may conclude that
influences, proceeding from the epiblast, can induce, some time
before fore-gut formation, the appearance of this structure from
non-presumptive endoderm; such influences probably operate in
the later primitive streak stages, with which we shall be concerned
in a later chapter. Other embryos of the same series, however,
exhibited phenomena which demonstrated the existence of an
influence proceeding in the opposite direction, from the hypoblast
and acting on the epiblast, at an earlier stage. In these blastoderms, the elongating primitive streak was bent away from its
presumptive direction so as to point towards the new position of
the anterior region of the hypoblast. As the embryonic axis
formed, the posterior part also became bent in the opposite
direction, so that the whole axis was shaped like a bow. An
example is sketched in Fig. 15 a. The embryos, which were
necessarily explanted into culture at a very early stage, could
not be kept alive long enough to develop very fully, but it
appeared that after the initial formation of the neural tube there
was a tendency for the embryo to become straightened out again,
and the curvatures lost. This straightening often began at the most
anterior end, and gradually progressed backwards along the body.

Still more striking evidence of the epigenetic influence of the hypoblast was obtained when the two layers were rotated through 180°, and replaced so that the hypoblastal axis pointed to 6 o'clock (Waddington, 1933 c). Seven different types of result were noted:

1. Embryo nearly normal, pointing more or less in the epiblastal direction.

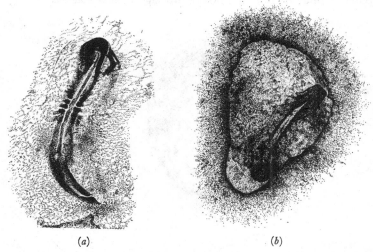

(a)   (b)

Fig. 15. a Duck embryo; endoderm of early streak stage placed at right angles to epiblast, with its anterior to right; curvature of both anterior and posterior ends of the axis towards the anterior of the endoderm. b Chick embryo: endoderm reversed in early streak stage (original epiblastal anterior towards top of page). Note the shortening of the embryo and its deviation from its original direction. (After Waddington.)

2 Similar, but embryo very short, or not stretching right across the area pellucida (Fig. 15 b).

3. Main part of the embryo pointing in a direction diverging widely from the epiblastal (Fig. 15 a).

4. Embryo in the epiblastal orientation, in which there had been a transitory appearance of a primitive streak pointing in the hypoblastal direction.

5. Semicircular embryos.

6. Two embryos, pointing in opposite directions, with their heads together (Fig. 16).

7. Single embryo in the hypoblastal direction.

In interpreting these results, it must be remembered that in most cases at the time of operation a short primitive streak was already present in the epiblast, which therefore had an inherent polarity and a tendency to form an embryo pointing to 12 o'clock; and since some embryos developed in this direction even when the operation was made before the appearance of the streak, this polarity must be initiated before any sign of it is visible.

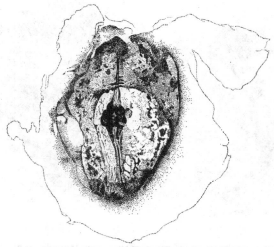

Fig. 16. Chick embryo; endoderm reversed in early streak stage (original epiblastal anterior towards top of page). A complete secondary axis has been induced, the two heads being united in a single complex formation about the middle of the blastoderm. (After Waddington.)

In class 1, the rearrangement of the two layers has not produced any great effect, although there is usually some differential inhibition of the posterior end of the developing embryo. The lack of effect is probably to be attributed to a failure in healing, to extensive injuries to the hypoblast during the operations, or to similar causes. In classes 6 and 7, on the other hand, it is clear that a new primitive streak has been induced and has developed into a new embryonic axis, while in class 4 there has been an induction which was later swamped by the developing epiblast. In the semicircular embryos of class 5, it is to be noted that it is always the posterior part of the axis which points in the hypoblastal direction, while the anterior part is in the epiblastal. This would seem to indicate that when a new primitive streak is induced, it is in the first instance the posterior part which is caused to appear. This is what might be expected *a priori*, since in normal development it is also the posterior region which forms first, the rest of the streak being built up by a streaming movement which proceeds towards the anterior.

It is clear from these experiments that shortly after it becomes a coherent layer, the hypoblast has a polarity which can influence the movements proceeding in the epiblast. In the 90° rotations, and possibly in the class 3 results of the complete reversals, it has diverted the movements out of their normal course; in the class 2 results it has inhibited them; and in classes 4, 5, 6 and 7 it has actually induced a new centre of movement which corresponds to the posterior end of a primitive streak. It will be remembered that in normal development the hypoblast probably takes part together with the epiblast in a polonaise-like streaming which involves a forward movement along the mid-line, and we saw that there is some reason to suspect that the hypoblast begins this movement earlier than the epiblast. Presumably it is some factor concerned in this movement which appears in these experiments as the hypoblastal polarity capable of inducing movements in the overlying epiblast. We shall find another example of the induction of morphogenetic movements at a later stage, when it is the primitive streak which has become the origin of dynamic tendencies (p. 102).

It seems probable, as was argued by Waddington (1933*c*), that the polarity is inherent in the hypoblast as a whole, as some form of 'gradient-field', rather than concentrated in a small specific area. If the latter were the case, we should expect that the active region of hypoblast would always either induce a new streak or have no effect at all; it would be more difficult to envisage the production of such results as the bending of the streaks in the 90° rotations. New results of Lutz, to be discussed below, conclusively demonstrate that the formative potentialities of the early blastoderm are not confined to any small region but are spread at least through the greater part of it. It is, of course, possible and indeed likely that there is some quantitative variation in the intensity of inductive capacity of the hypoblast, just as there is, for instance, a variation in the tendency to autonomous movement in the roof of the amphibian gastrula in the neighbourhood of the blastopore.

We have so far attributed the inductive effect to the hypoblast, and it remains to consider the possibility that it is rather to be attributed to the few mesoderm cells which may adhere to the lower layer when that is separated from the upper. This was discussed by Waddington (1933c) and seems very unlikely for several reasons. In the first place, very little mesoderm actually adheres to the hypoblast. Rudnick (1944) has suggested that even a few cells might be enough to induce the initiation of a new centre of invagination; but they could hardly be enough to exert such widespread influence as is exhibited by the bending of the streaks in the 90° rotations. Moreover, the inductive action is most often successful in the earliest stages, when little or no mesoderm has been formed. It is, of course, scarcely ever possible to be completely certain that an experimental isolate contains absolutely no cells of a kind different to that intended, and it is, therefore, always theoretically possible to attribute any action the isolate may show to an unverifiable admixture of foreign cells; but in this case the hypoblasts seem to consist as purely of endoderm as can reasonably be demanded, and there would appear to be no good reason for attributing their inductive influence to anything else.

However, although it is probably safe to attribute the epigenetic effect of the reversed hypoblasts to their endodermal component, that does not by any means rule out the possibility that the epiblast may, at the same stage, also possess an effective polarity. It has been mentioned that most of the experiments on reversal of the hypoblast were actually performed on embryos in which the primitive streak had just appeared, and in these it is certain that the epiblast can no longer be considered a mere neutral sheet on which the hypoblast can exert its influence. There is, therefore, nothing in the experiments which would make it impossible to suppose that at any earlier stage, before the appearance of the streak, the epiblast might already have some similar intrinsic properties. It might be that the polarity is common to the two layers, perhaps originating before the process of delamination splits up the original mass of morula cells; if this were the case, the induction by the hypoblast which can be demonstrated in experiment may have no essential part to play in normal development.

Separation of epiblast and hypoblast

Some evidence as to the properties of the early epiblast can be obtained from attempts to cultivate the two layers of the didermic blastoderm in isolation. We may first mention shortly the performance of the hypoblast. Neither *in vitro* (Waddington, 1932) nor on the chorio-allantoic membrane (Hunt, 1937 a) has it proved capable of differentiating into characteristic endodermal organs, if taken from stages younger than the definitive streak. However, this failure is most probably to be attributed mainly to the abnormal mechanical conditions to which it is subjected; at least *in vitro* it can be seen that young isolated hypoblasts become crumpled up into a distorted mass, and the conditions on the chorio-allantois are likely to be even worse.

The *in vitro* technique seems also to be unfavourable to the development of young isolated epiblasts, although older ones from middle and late streak stages develop quite well. There is always a tendency for epithelia *in vitro* to form cysts, and a transformation of this kind usually occurs rather strongly in epiblasts isolated before the appearance of much mesoderm; the cysts take the form of 'bubbles' appearing all over the area pellucida. Differentiation of epiblasts from pre-streak stages was in consequence never observed. Spratt (1946, p. 284) also found, in a small experiment, that when the hypoblast was removed in a pre-streak stage and the epiblast cultivated *in vitro* no primitive streak developed. Slightly older stages, which were nevertheless young enough to be influenced by the hypoblast in the ways described above, could develop comparatively well-formed axes containing the usual ectodermal and mesodermal organs (Waddington, 1932). These axes were always considerably shorter than normal, and it may be that the normal elongation of the primitive streak is inhibited or even ceases altogether when the epiblast is separated from the hypoblast at an early stage; but it must be remembered that the formation of short embryos is a common result of the conditions of *in vitro* culture and too much significance should not be attached to it in particular cases.

On the chorio-allantois, young epiblasts develop better. Dalton (1935) claimed that, when the hypoblast was removed from the region of the anterior end of the young streak, mesoderm formation ceased, and that the same was true till an even later stage at more posterior levels. Hunt (1937a), however, obtained considerable developments from epiblasts separated before the appearance of the streak, so that this negative evidence must be rejected. In Hunt's grafts, the epiblasts gave rise to many ectodermal and mesodermal tissues, such as central nervous system, skin, feather germs, cartilage, mesonephros, heart, etc. They also, in spite of the absence of hypoblast, developed into some endodermal tissues, such as gut, stomach, liver, etc. This apparently *bedeutungsfremde* differentiation is discussed elsewhere (p. 92). In the present context, the important point is that the epiblast even from the pre-streak stage is capable of considerable development in isolation. Therefore, it is certainly not a merely indifferent tissue, awaiting the inductive stimulus of the hypoblast. On the other hand, it is equally certainly not rigidly determined, since a new embryonic axis can be induced in it. It is, then, in a state of labile determination which presumably begins as a weak tendency to primitive streak formation and gradually becomes intensified until the streak is fully formed and the axial organs begin to appear. We shall see later that the presumptive ectodermal parts of the epiblast retain their full responsiveness to inductive stimuli until the end of the streak stage, so that it is probably mainly the presumptive mesoderm (and perhaps the pre-chordal neural plate) which begins to be determined in the pre-streak stage.

Isolation experiments in the pre-streak blastoderm

As might be expected, a considerable number of experiments have been performed on the early blastoderm which did not involve the separation of epi- and hypoblast, but rather studied the developmental capacities of different areas of the blastoderm, or its response to localised injury. We may first consider the extensive series of chorio-allantoic grafts of portions of pre-streak or early streak blastoderms. Unfortunately, most of the theoretical arguments which have been derived from this work have been based on negative results. It has usually been found (with a few exceptions noted below) that any region which differentiates at all gives rise to most of the tissues of the young embryo, so that rather little can be deduced from the nature of the differentiations obtained. Several authors have, therefore, attempted to obtain a deeper understanding of the epigenesis of the blastoderm by noting which areas cannot differentiate. But there is always the possibility that such failures of development may be due to the inadequacy of the conditions in which the isolate was cultivated; and, in fact, later work has only too often shown that the negative conclusions drawn by earlier workers were not firmly based.

Thus most of the earliest workers with the chorio-allantoic method (Danchakoff, 1924; Hoadley, 1926a; Umanski, 1929) failed to obtain any noteworthy differentiation from grafts of the entire unincubated blastoderm; and on this basis, coupled with the fact that good development occurred from slightly later stages, Hoadley erected a theory of 'segregation' or 'differential dichotomy' in which he discussed the gradual acquisition, by different regions of the embryo, of a capacity for self-differentiation. However, in the hands of Murray & Selby (1930a) even pre-streak stages would differentiate on the chorio-allantois, as they also did in later experiments by Umanski (1931). These workers also made grafts of parts of the blastoderm. Murray & Selby found less differentiation from transverse thirds and fourths than from whole blastoderms, and cautiously attributed this to the inhibiting effect of the operation. Umanski (1931) found some development of neural and mesodermal tissues from posterior halves, but not from anterior. Similar results were obtained by Butler (1935). This author fragmented the unincubated blastoderms in a large number of different ways. In grafts of the whole blastoderm, neural tissue developed in a rather small number (some eight or nine), while recognisable axial mesoderm (such as chorda or cartilage) was observed in about five specimens. Similar results were obtained with some grafted fragments. Neural tissue was found in some twenty-three grafts, derived from various different regions. Axial mesoderm occurred much less frequently, and Butler claimed that it is only produced from fragments which contain the uninjured posterior median quadrant, that is, the presumptive streak area, which is stated to be able to produce all the organs of the adult. This it may well be able to do, but recognisable chorda was found only five times, brain tissue three times, eye tissues three times. In face of these low figures for positive results, it is difficult to take very seriously any argument based on the failure of other regions to develop into similar tissues in grafts. Moreover, Hoadley (1926a, b) had already obtained some differentiation from anterior strips of the blastoderm, aged only a few hours, both on the chorio-allantois, and in embryos sectioned *in situ* in the egg. We shall see later that there is convincing evidence of a positive nature that parts of the blastoderm isolated from the presumptive streak area are well able to differentiate under suitable conditions.

From those parts of the blastoderm which have, in the hands of various workers, failed to give neural tissue or axial mesoderm, it has been usual to obtain the differentiation of epithelial tubules, gut and smooth muscle. Similar muscular tissue was also found by Olivo (1928) in tissue cultures of the edges of unincubated blastoderms. He interpreted it as heart muscle, largely because it acquires a rhythmical beat, but such an activity seems to be characteristic of smooth muscle *in vitro* and cannot safely be used to identify the muscle as belonging to the heart.

A thorough study of the development of parts of the young blastoderm cultivated by the watch-glass method *in vitro* has been made by Spratt (1942), who obtained more consistent results than the workers with the chorio-allantoic method (Fig. 17). From the pre-streak blastoderm, small pieces of neural tissue, which he interpreted as forebrain, developed from the anterior half in rather over 50 % of the explants. If the cut was made further towards the posterior end, separating the anterior two-thirds from the posterior third, neural tissue developed in nearly all the anterior pieces and in a majority of the posterior ones, while chorda was found in equal frequency (about 15 %) in both. When the cut was still further back, the small posterior fragment gave neural tissue in a lower percentage, and yielded no chorda; the fragment in this case would have contained none of the normal presumptive areas for these tissues. Essentially similar results were obtained in early streak stages.

Rather similar experiments had been made a few years earlier by Rudnick (1938c); she cut the blastoderm in the short primitive streak stage into three pieces by V-shaped cuts which were intended to separate the presumptive neural and mesodermal areas but which certainly would not separate them if Spratt's ideas as to the presumptive areas are justified, nor according to any current ideas. The fragments were cultivated *in vitro*. Unfortunately, recognisable axial mesoderm developed very rarely from these early stages (one doubtful case of chorda). On the other hand, all areas gave rise to mesenchyme, though only the posterior pieces gave twitching muscle (heart ?), and these also gave a much higher proportion of red blood cells. Neural differentiation was again comparatively rare. Its distribution between the areas does not seem to allow of any firm statement that neural tissue can, from these stages, develop in the absence

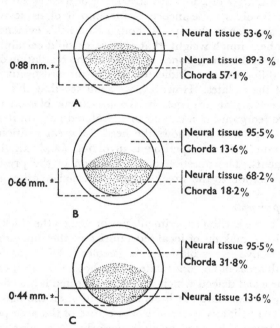

0·88 mm. ±

Neural tissue 53·6%

Neural tissue 89·3 %
Chorda 57·1%

A

Neural tissue 95·5%
Chorda 13·6%

0·66 mm. ±

Neural tissue 68·2%
Chorda 18·2%

B

Neural tissue 95·5%
Chorda 31·8%

0·44 mm. ±

Neural tissue 13·6%

C

Fig. 17. Tissues differentiating in cultures *in vitro* of parts of the pre-streak blastoderm, transected at various levels. (After Spratt.)

of mesoderm (in spite of the confident statement of this in Rudnick, 1944, p. 201).

Both Rudnick and Spratt interpret their results as showing that when parts of the pre-streak and early streak blastoderm are isolated *in vitro* they differentiate into those tissues whose presumptive areas were included. This conclusion, however, is by no means certain as regards neural tissue, since it seems doubtful if the main mass of this was ever satisfactorily separated from presumptive mesoderm. But we shall see that there is some evidence suggesting a capacity for independent differentiation by the most anterior, pre-chordal neural region in later stages, and it is possible that such a tendency has been exhibited, for instance, in Spratt's anterior pieces. Both authors also conclude that the very early streak does not extend forwards to the presumptive chordal region, which would imply that the elongation of the streak is not entirely a matter of a forward streaming of its original material, but also involves some incorporation into it of tissue which at first lies anterior to the tip. Although Rudnick's evidence is too slight to bear much weight, that of Spratt would certainly imply something of this kind, if one supposed that no regulation occurred and his differentiations indicated strictly the prospective significance of the isolates. However, we shall see that this is by no means certain; at any rate in the pre-streak blastoderm very extensive reorganisation can occur in isolated parts of the blastoderm (p. 64), and this shows the need for great caution in the interpretation of any experiments involving simply the isolation of fragments. The question of the location of the presumptive areas in relation to the streak is best discussed on other grounds.

Injury experiments

Some of the earliest experiments performed on the chick embryo involved the making of localised injuries in the unincubated or early primitive streak blastoderm. Peebles (1898, 1904) claimed that small areas of necrosis made with a hot needle in these stages led to the exact defects which would be expected if the embryonic body was already laid out in its final position on the blastoderm, with its head slightly anterior to the centre of the area pellucida and the trunk stretching away posteriorly. A similar view was expressed by Assheton (1896) on the basis of marking various regions by the insertion of sable hairs. Kopsch, both in the early days and also much more recently (1926, 1934) has supported the same thesis, on the basis of the defects produced by killing small areas of the early blastoderm with an electrolytic device. Such conclusions are, of course, in complete contradiction with the maps of presumptive areas based on vital staining, cinema-photography and the *in vitro* experiments. But, in the first place, the interpretation of Peebles and Kopsch overlooks the possibility, which must be envisaged nowadays, that a defect in an ectodermal structure may be produced not only by a direct injury to it, but also by an injury to the underlying mesoderm which induces it.

Moreover, many of the embryos which develop after these defects are highly distorted, owing to the tendency for the pressure of the yolk to enlarge any hole made in the blastoderm, and their interpretation is thus often somewhat obscure. Kopsch's later embryos, in particular, were not always very thoroughly examined, and Twiesselmann (1935, 1938), who repeated many of Kopsch's experiments, showed that embryos, which superficially appeared very similar to those which Kopsch claimed to show only localised defects, in many cases really were much more abnormal than appeared at first sight. Little reliance can, therefore, be placed on the results described by Peebles and Kopsch.

Twiesselmann was also able to show that after injury the early blastoderm does not always develop as a mosaic of determined parts, but on the contrary has very considerable powers of regulation. Localised injuries, by electrolytic needle or ultra-violet 'Strahlenstich' may expand into considerable wounds which almost divide the blastoderm into two parts, and the same effect is, of course, produced by cutting the germ with a knife; and, in such cases, more than one complete embryo may develop. Twiesselmann, in fact, described a few examples of two relatively well-developed and complete embryonic axes, and some in which three less complete structures were present. These embryos were usually more or less distorted, by the effect of the yolk in tearing apart the edges of the wound.

A few years before Twiesselmann's work, Morita (1936, 1937) described a series of experiments in which killed adult tissues were placed in contact with the unincubated blastoderm, and resulted in the appearance of very perfect double or multiple embryos. He attributed this to the action of evocators diffusing out of the dead tissues, and compared the results with those which had become well known in the Amphibia. When he proceeded to find that a similar effect could be produced by inert substances, he adopted the hypothesis, which had first been advanced by Waddington, Needham & Brachet (1936) for the Amphibia, that agents which kill a certain number of cells in a localised area cause the release from them of evocating substances to which the remaining healthy cells may react. But Morita's supernumerary embryos are very unlike anything else which has been produced by evocation in the avian embryos; and although it is true that some weight must be given to the fact that he operated on an earlier stage than that used in studies of induction by the primitive streak, it appears most probable that his results are due not to evocation in the usual sense, but to a reorganisation following partial separation of the germ into isolated areas.

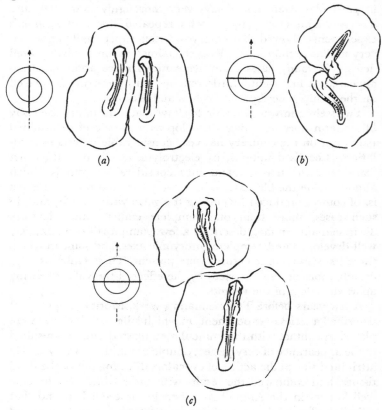

(a)

(b)

(c)

Fig. 18. Twins developed after transection of the unincubated blastoderm. *a* Presumed longitudinal section; *b, c* presumed transverse sections. Note reversed orientation of anterior embryo in *b*. (After Lutz.)

More recently, Lutz (1948 *a, b, c, d*; see also Wolff, 1948) has obtained very striking results of a similar kind in the duck. In the latter species, the egg is laid at a rather earlier stage than that of the hen, so that in Lutz's experiments, the formation of endoderm is usually only just beginning; moreover, the distorting effect of the yolk is very much less. The operations consisted in cutting the blastoderm into a series of strips with a glass needle; some of these might then be eliminated by pushing them down into the yolk, where no further development seemed to take place. Unfortunately, the orientation of the blastoderm cannot be made out with certainty at this stage; one has no alternative but to rely on von Baer's rule, which Lutz showed was true in about 65 % of his eggs. The beautiful multiple embryos obtained by Lutz are important in several different ways. In the first place, they show that almost, if not quite, every part of the unincubated blastoderm is capable of developing the axial organs of the embryo. The restriction of this capacity to the posterior quadrant, found by the workers with the chorio-allantoic membrane (cf. p. 59), does not hold in these different conditions of experiment. Clearly, on the yolk the fragments of blastoderm are capable of great regulation, or paragenesis, in Dalcq's terminology, whereas on the chorio-allantois or in tissue culture they develop only into their presumptive fate (a normogenesis) or even less.

Again, it is noteworthy that the embryos which develop in the separate strips are all complete or nearly so; the paragenesis is 'teleogenetic', in that it leads to the formation of the normal embryo.

The orientation of the embryos is also of extreme interest, but considerable obscurity (Lutz, 1950 a, b, c). When the line of section is parallel to the anterior-posterior axis of the blastoderm, the twin embryos which develop point in the same direction, to 12 o'clock (in our conventional terminology, see p. 52). The same is often true when the cut is perpendicular to the axis, but in such experiments the anterior embryo may have its polarity reversed, pointing to 6 o'clock, so that the two embryos lie head to head. This happens the more often, the earlier during incubation the operation is performed. Moreover the reversed orientation is more frequent when the line of section lies anterior to the centre of the blastoderm. Several other possibilities of orientation have been realised. The interpretation of these results is still obscure. We presumably have to do with the reorganisation of parts of a field, which originally extends throughout the whole blastoderm. Such an extensive field is similar to that postulated by Waddington (1933 a) on the evidence of the widespread effects of endoderm reversal, such as the bending of primitive streaks after 90° rotations (see p. 53). It is tempting to try to explain Lutz's results in terms of the sequence of events during endoderm formation. We know (cf. Peter, 1938a) that this layer first forms round the edge of the blastoderm, then spreads into the posterior region and from there over the whole under-surface. One might on these grounds be tempted to postulate a field of 'endodermal potential' something like that indicated in Fig. 19 in which the endoderm of the streak region has been granted a higher potential than the non-axial endoderm. Such a hypothesis would enable one to bring order

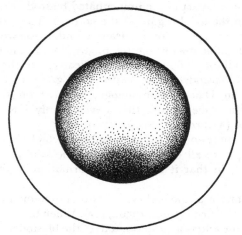

Fig. 19. Diagram of a hypothetical field of 'endoderm potential'.

into some or Lutz's unexpected results, such as the reversal of polarity in anterior pieces.

Lutz has also described a number of experiments in which an incision was made in the direction of the axis but stretching only part of the way across the blastoderm. If the cut lay in the anterior radius, the result was a double-headed embryo (Fig. 20); if it lay in the posterior, he found, similarly, a duplicitas posterior. The length of the duplicity varied greatly, even in embryos which were thought to have 'been submitted to cuts of the same length; in some cases a partial cut has led to the development of two completely separate embryos. From the latter result, we may conclude that even an incomplete division of the blastodermal field may cause it to become reorganised into two distinct and complete units, a phenomenon which, as will be seen (p. 72), may also occur at a later stage. But apart from this, it seems to be difficult to draw any firm conclusions from the appearance of the anterior and posterior duplications. The great variation in the length of the axis which is doubled could be explained as a result of the forward and backward movements along the streak which

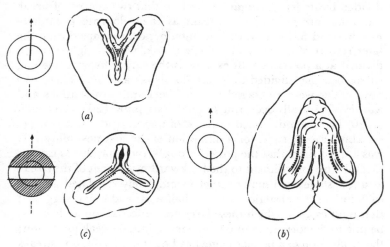

(a)

(c)   (b)

Fig. 20. *a, b* Duplicitas anterior and posterior following partial transections of the blastoderm; *c* twin embryos, with partially fused heads, from a transverse strip of blastoderm, the remainder of which was eliminated by immersion in the yolk. (After Lutz.)

we have seen to be likely; but it would also be expected if we granted merely that the streak increased in length by intra-susceptive growth, in the manner suggested by Jacobson and Spratt (see p. 21). Similarly, the fact that, when an anterior cut is made, the double heads extend well forward of its posterior end (and the corresponding fact about the posterior duplications) certainly argues for the progression of something in an anterior direction in the growing streak (or in a posterior direction at a later stage); but it says nothing as to whether this progression is an actual movement of tissue, or whether it is a wave of streak-formation which moves forward through a stationary tissue.

But although the duplications do not offer any conclusive evidence on the vexed questions of the morphogenetic movements during the primitive streak period, they seem easily explicable in terms of the processes which we have, on other grounds, held to be likely. An anterior duplication can easily be supposed to arise when the forward movement in a growing streak divides in two on reaching the region in which the blastoderm has been split. A posterior duplication could be formed either when a posteriorly divided blastoderm reorganises itself so that two centres of streak formation are set up, the streaks as they elongate eventually meeting and fusing towards the anterior; or, perhaps during the later retreat of an originally single streak, which might become divided at a posterior split exactly as the earlier forward-moving streak becomes divided to give a duplicitas anterior. These two possibilities were envisaged well before Lutz's observations and a series of naturally occurring duplications, published by Newman (1940) had been discussed in terms of them (Waddington, 1941; see also Gräper, 1931, for a discussion of the first possibility). It was argued there that the posterior duplications produced experimentally by impediment to the backward movement of the streak at a late stage are usually rather incompletely regulated, so that each fork of the posterior is only a half-axis; and it was suggested that most cases in which these forks are perfect are more likely to be due to the formation of two regions of initial streak formation; the production of a perfect single head by the two colliding streaks was not held to be improbable, since regulation is much better at the early stage at which this would occur. Lutz's posterior duplications, resulting from cuts along the posterior radius in the unincubated blastoderm, are almost certainly produced in this way. But the first of the arguments above, that regulation is usually incomplete at later stages, no longer appears so strong, in view of the results of Abercrombie & Bellairs (1951, p. 72). It is possible, therefore, that some naturally occurring posterior duplications do not arise until the period of the retreating streak.

CHAPTER V

DEFECT AND ISOLATION EXPERIMENTS ON THE PRIMITIVE-STREAK STAGE

ONE method of investigating the epigenetic situation during the period just before and after the appearance of the axial organs is the broad category of what may be called injury experiments. These may be of various kinds. The developing area, whether it is the whole blastoderm in the primitive streak or head process stages, or the smaller region to which the streak is confined at later stages, may be simply cut into two pieces, which may then be allowed to develop *in situ* on the yolk, *in vitro*, or as chorio-allantoic or coelomic grafts. The fragmentation can be carried further, and the area cut into three or more strips, which subsequently develop in the same ways. Or even smaller regions may be isolated and their development studied. And any of these fragments may be cultured with or without the hypoblast, although unfortunately the close adhesion of the invaginated mesoderm to the ectoderm, and the lack of precision in our maps of the presumptive areas, makes it impossible to isolate fragments which are known to be homogeneous in respect of their presumptive fate. Finally, the obverse of the isolation of small fragments is the observation of the developmental effects of small defects made to the blastoderm, either by the removal of a piece or by killing a group of cells in some way, such as by the use of X-rays (Wolff, 1936) or ultra-violet (Heinrichs, 1931).

By far the greater part of the experimental work which has been done on the chick falls in this general category. Before discussing the results in detail, it will be as well to mention some of the respects in which particular caution is necessary in their interpretation. An isolated part of an embryo or an embryo in which a certain region is deficient may develop exactly into its presumptive fate; it may undergo, in Dalcq's phrase (1941; cf. Raven, 1948), a normogenesis. But we must be prepared for its development to diverge from what it would have been in the uninjured whole. Such a changed course of development has been spoken of by Dalcq as a 'paragenesis'; it may be of such a kind as to lead towards the normal unitary organism, in which case it is called 'teleogenetic', or it may tend towards some other end-state, that is, it may be ateleogenetic. The word 'regulation' is sometimes used in the sense of paragenesis of any kind, sometimes only of teleogenetic paragenesis; it is, therefore, not very suitable for exact usage, although it may still serve a useful purpose in cases where precision in this respect is not of the first importance.

In injury experiments on the chick, we must expect still further complications caused by the conditions under which the operated embryos or fragments are cultivated. The general injury effects of the necessary cuts may themselves affect the developmental capacities of the pieces; the chemical properties of the isolation medium might tend to a general suppression of development, or it might, owing, for instance, to the presence of an evocator substance, tend to stimulate differentiation of a particular kind; the mechanical conditions, by furthering or inhibiting certain morphogenetic processes, might tend to facilitate or impede development; and, finally, the culture conditions may encourage normogenesis or some kind of paragenesis. Therefore we must not be surprised if we find that in a variety of experimental conditions different courses of development are pursued. These considerations would apply even to fragments which consisted wholly of one type of tissue. They apply even more strongly to most of the chick isolates, which usually include at least presumptive ectoderm and presumptive mesoderm, and often endoderm as well.

Thus in discussing the injury experiments we shall always be on rather insecure ground. They can tell us little about the processes of normal development except when a part develops into what we should on other grounds have expected to be its presumptive fate; such an apparent normogenesis has a certain confirmatory value. We shall also find ourselves confronted with a series of interesting facts concerning paragenesis—its extent under different circumstances, its variation with time and so on. From these facts we can, with due caution, make some deductions as to the degree which specific local differentiation has attained at various times; but the facts will unfortunately do no more at present than set rather wide limits to speculations as to the nature of the developmental processes.

Defect experiments

It will perhaps be as well to start our consideration of the injury experiments by an account of the effects produced by small defects made to the primitive streak blastoderm. If a small hole is made in the mid-line at this stage, by the removal or killing of a piece of tissue, there is a considerable tendency for the hole to become enlarged. This is presumably partly due to the effect of the longitudinal movements which occur during streak regression, but in blastoderms operated in the egg it is also encouraged by the pressure of the yolk. The tendency is, therefore, least noticeable in more stable mechanical conditions such as cultivation on a solid plasma clot *in vitro*. In these circumstances the hole may be completely closed up, and regulation to a normal embryo may take place (Waddington, 1932). It is even possible to obtain an embryo which is complete (at least in respect of the anterior segments which develop during the comparatively short life of an *in vitro* culture) when the whole primitive streak, and a strip of moderate width on each side, have been removed. The filling up of the hole made by the defect, which is the first step in the restitution of normality, seems to occur always under the influence of the transverse movements connected with the invagination of mesoderm; if the wound is still open when the longitudinal regression movements reach that level, instead of assisting the closure, they tear it further open. A similar closure by transverse movement, followed by regulation to the normal shape, can occur in front of the node, in the region occupied by the presumptive brain and floor of the neural tube. The regenerative phenomena in this region are discussed in more detail on p. 140.

Regulation in these stages is dependent on the previous restoration of continuity of tissue. If the original wound is not closed, but gapes into a wide hole, neural tissue and somitic mesoderm may be found along each of the lateral margins of the space. They usually represent only the lateral walls of the nervous system and rarely become reconstituted to form a complete embryonic axis on each side, or even undergo enough regulation to give rise to notochord unless the presumptive area of that tissue was left. Waterman (1936) states that in some cases in which a sagittal cut was made along the mid-line in the anterior end of the streak, two more or less symmetrical heads were formed. But in general, when a region of the embryo develops along the margin of the sheet of tissue, it shows a strong tendency to do so in a mosaic manner. In striking contrast to this the same areas of presumptive tissues may not only regulate to give one complete body if the hole closes in time, but may regulate to form two complete axes if the wound is filled by some indifferent tissue. Abercrombie & Bellairs (1951) removed small pieces containing the node from the definitive streak stage, essentially similar to the pieces which had been reconstituted in Waddington's experiments; they filled up

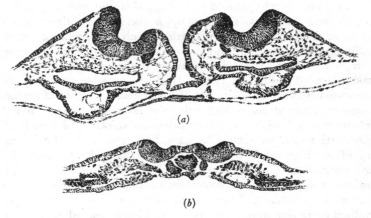

(a)

(b)

Fig. 21. Sections through a duplicitas produced by substitution of a graft of presumptive epidermis for the node region at the primitive streak stage. (From an unpublished photograph of Abercrombie.)

the hole by inserting a suitably sized piece of epiblast. The result in some cases was the production of two complete embryonic axes. Clearly the presumptive material left on each side had reorganised itself into a whole unit (Fig. 21).

The production of two complete embryos by reorganisation of one gastrula is, of course, well known in the Amphibia; and we shall see that, epigenetically, the primitive streak stage of the chick is in much the same condition as the young amphibian gastrula, that is, it is just entering on the phase when the axial organs become finally determined by an organiser action. Further, the dependence of the degree of reorganisation achieved on the restoration of topographical continuity is also a well-known phenomenon in that group (see discussion by Lehmann, 1945, p. 246).

In fact, Bautzmann (1932) has described experiments in *Triton* which are almost exactly the same as those of Abercrombie & Bellairs in the chick. He found that, if a piece of ventral presumptive mesoderm (*Randzone*) is transplanted in place of an equivalent piece of dorsal presumptive chorda, considerable duplication of the axes may occur, ranging from duplicitas anterior to cases with two complete axes, each of which by reorganisation has become fully symmetrical. The chick regulation is, therefore, no more than can be found in the Amphibia.

When injuries are made to the chick blastoderm *in situ* on the yolk, little sign of restoration or reconstitution can be found, probably because the mechanical conditions tend to cause the wound to open out into a large hole. The early experiments of Peebles (1898, 1904) showed that the node region could develop into the head and a few segments of the body when the posterior part of the streak was destroyed, but she found no development of any part in blastoderms in which the node itself had been eliminated. Wetzel (1924) obtained essentially similar results, and at first concluded that the node acts as an organiser, in the absence of which the other parts of the embryo could not develop. But further investigation led him (1926) to withdraw this (cf. also Umanski, 1928). By 1929 he had found (1929b) that if a cut is made immediately anterior to the node, in the definite streak stage, a little brain tissue (presumably fore-brain) can develop in front of the cut, while the rest of the embryo is formed more or less normally behind it. If the section lies just posterior to the node, axial organs develop from the anterior part, but the hind part forms only at best somitic mesoderm, with no neural tissue, the hole always enlarging into a large space, owing to the absence of the organs of the mid-line, the chorda and floor of the neural tube. If Wetzel eliminated the node, he also found that the hole enlarged, and the axial organs were missing. He claimed that these results fitted in well with the disposition of the presumptive areas, although one might have expected to find presumptive neural tissue extending some way posterior to the node; and Wetzel could at that time still speak hesitatingly of 'eine gewisse Regulationsfähigkeit dieses eigentlich idealen Mosaikeies', a description which does not seem very adequate in the light of the regenerative phenomena mentioned a few paragraphs earlier in this account.

Wolff (1936, in which references are given to many preliminary notes) has published a large study on the results of local defects made to blastoderms *in ovo*. In the majority of experiments the defects were made by irradiation with X-rays, the greater part of the blastoderm being protected by lead shields, so that only a localised spot of tissue was affected. The exact nature of the irradiation effect is not precisely described, but it appears that usually the irradiated tissues take no further part in differentiation; however, they do not disappear so as to leave an open wound.

Unfortunately Wolff's point of view in planning his experiments was not that of an experimental embryologist, but that of a 'teratologist'. That is to say, he set himself the aim, not so much of understanding the causal processes of development, but of finding experimental means of producing certain definite types of abnormalities, or 'monsters', which have been recognised and described under a vast series of Greco-Latin names by the older anatomists. He was thus tempted to keep his embryos alive so long after the operation that very many secondary processes may have intervened to complicate the interpretation of the results; and he devoted an extraordinary amount of time and trouble to the production of types, such as omphalocephalic, cyclopic, or symmelic monsters by methods which indeed achieved their end, but give extremely little insight into the course of normal development.

Irradiation of the region of Hensen's node, in the fully formed streak stage, always led to some deficiency in the embryonic axis (Wolff, 1934*a*, 1935). This usually took the form of the appearance, along a section of the body somewhere in the neighbourhood of the tenth segment, of a series of single unpaired somites instead of the usual paired ones. Waddington (1932) had already described two examples of the formation of a series of unpaired somites in embryos from which Hensen's node had been removed and the wound incompletely regulated. In Wolff's specimens, normal somites were present anterior to the defective region, and might also be found posterior to it. The notochord is absent in the region of unpaired somites, and apparently also posterior to it, even if the posterior region contains paired somites. The neural tube may also be defective and, if so, the abnormal region always corresponds exactly with the abnormal somites. The latter

observation suggests that the deficiencies in the neural system
are not produced directly, but are consequences of an abnormal
inductive action by the injured mesoderm; it is well known that
injuries to the invaginating mesoderm in Amphibia may produce
deficiencies in the overlying neural system, although the latter
remains perfectly healthy (Lehmann, 1926).

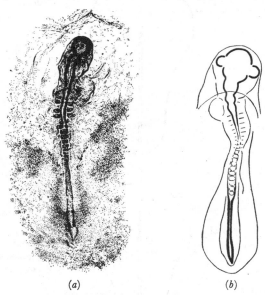

(a)   (b)

Fig. 22. *a* Embryo developed after removal of Hensen's node from a long
primitive streak stage. Considerable regeneration has occurred, but the
posterior somites are fused in the mid-line. (After Waddington.) *b* Fused
posterior somites in an embryo developed after X-irradiation of the node
region. (After Wolff.)

Irradiation just anterior to the node may produce a complete
abortion of the head (Wolff, 1936, p. 308). In other cases (pp. 97
et seq.), probably with a lighter irradiation, the effect was merely
to produce an 'omphalocephalic monster', i.e. an embryo in
which the folding of the head fold and foregut was completely
abnormal (see p. 104). Posterior irradiations produced some
well-defined deficiencies which on the whole agree well with what
might be expected on the basis of the map of presumptive areas.
There seem to have been only two experiments (pp. 301–7) which

affected the region immediately posterior to the node. One, whose anterior border was 0·3 mm. posterior to the primitive pit, gave rise to an embryo which was normal nearly to the level of

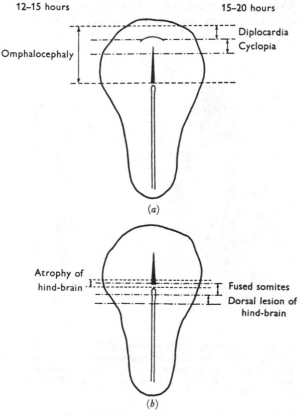

Fig. 23. Effects produced by X-irradiation of narrow transverse bands of the blastoderm at (a) the streak and (b) head process stages. The positions of the head process and head fold are indicated in a, although they would not be present at the time the irradiations were made. (After Wolff.)

the posterior limbs, but completely lacked a tail bud; it is worthy of note that, as the neural tube is traced posteriorly, it is its dorsal side which first becomes defective. In the other embryo, the zone of irradiation extended forward just to the level of the node, and the neural tube was nearly completely aborted posterior to the neck region, although the notochord was normal back to the hind-limb area. Irradiation of still more posterior parts of the streak gave defects of the mesoderm in posterior parts of the body. In some of these cases there seem at first sight to be more un-injured somites anterior to the defective region than would be found in embryos similarly operated *in vitro*; but this may be a consequence, not of any regenerative or regulative phenomena (which, of course, remain possible), but simply of the fact that Wolff's embryos could be kept alive longer, and could thus develop further than the *in vitro* ones.

Transections of the blastoderm, and isolation experiments

Wetzel's original claim that the node was essential for further development, from which he seems at least partially to have withdrawn by the time of his note (1929b), received full support from some of the American workers with the method of chorio-allantoic grafting. Hunt (1931a) and Willier & Rawles (1931b) both found that posterior portions of the streak-stage blastoderm, isolated from the node, could not develop any neural tissue, chorda or other tissues characteristic of the embryonic axis, whereas the node level formed all the axial organs, at least those occurring between the anterior and the mesonephros. From the negative evidence of failure to obtain differentiation in post-nodal isolates, they concluded that the node is an organiser in Spemann's sense. The logic of this deduction is obviously faulty, and betrays a misunderstanding of what an organiser is, a point which American authors, who had developed a peculiar embryological theory of their own in terms of 'embryonic segregation', often seemed to find difficulty in grasping (see p. 231). But perhaps more important is the question of fact; is the node actually essential for the development of the posterior regions?

As a matter of fact, evidence of the dispensability of the node, derived from experiments *in vitro*, had already been published in a preliminary way (Waddington, 1930). In the later and fuller pub-lication (1932) a number of experiments were described in which blastoderms of the streak stage were cut in two at various levels along the axis, and the parts cultivated. As would be expected in the absence of regulation, no axial organs developed on anterior pieces cut some distance in front of the node. If the cut was made through the node itself, a little neural tissue (presumably fore-brain) appeared on the anterior part, and a nearly or quite complete embryo on the posterior. Such anterior pieces can also form fore-brain on the chorio-allantois (Stein, 1933). Recent, more exact, study of their behaviour *in vitro* (Spratt, 1942) has shown that the potential fore-brain-forming region extends about 0·3 mm. anterior to the node; if the cut is made further forward than that, no neural tissue appears on the anterior piece; thus this limit probably does not attain the anterior margin of the presumptive brain area (Fig. 24).

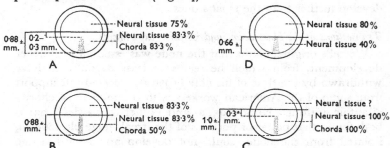

Fig. 24. Tissues differentiating in cultures *in vitro* of fragments of primitive streak blastoderm transected at various levels. Upper row, short streak; lower row, medium streak. The anterior piece forms neural tissue in all cases except that of a medium streak stage transected some distance anterior to the tip of the streak (bottom right). (After Spratt.)

Most interest in the present context attaches to cases in which the cut was made some small distance posterior to the node. On the anterior parts in such experiments, there developed a head which extended posteriorly into a long thin process, which usually projected well beyond the posterior margin of the piece, and was made up of the main axial organs, neural plate, chorda and somites (Fig. 25 a). The tendency to 'tail' formation becomes less as the cut is further removed from the streak towards the posterior, and it also falls off if the operation is made at later stages. The formation of such projecting 'tails' seems to present clear evidence for a real backward migration of tissue during the regression of the streak, and it is difficult to believe that this is a regulative para-genesis which does not occur in normal development. If the cut is made obliquely, a half axis may be formed running down the side which stretches towards the posterior; presumably the backward-streaming tissue has been able to find a passage along that edge

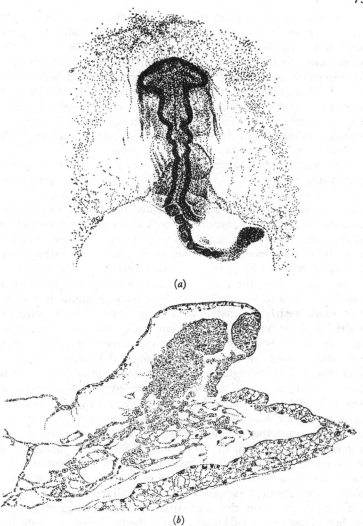

(a)

(b)

Fig. 25. *a* The second quarter (counting from the anterior towards the posterior) was removed from a fully grown streak, and the hole became enlarged. The anterior quarter of the streak has formed a head and a 'tail' extending backwards into the hole. (After Waddington.) *b* Section through one limb of the V-shaped embryo developed from a 22-hour blastoderm sectioned very slightly posterior to the node. (After Waddington.)

of the cut, and it seems in this way to reach into an originally lateral region of the blastoderm in which normally no axial structures would be developed. The phenomenon is in fact exactly what would be expected if the fragment developed into its presumptive fate according to the suggestions which had been made by Wetzel, and it seems strong evidence against the later ideas of Jacobson (1938b). Similar though less well developed 'tails' were later figured by Wetzel (1936), after operations *in ovo*; and Rudnick (1938b) and Spratt (1947c), repeating Waddington's experiments, obtained exactly the same result.

If the cut is only a fairly short distance behind the node, some embryonic structures also develop in the posterior part. As would be expected in a normogenesis based on the presumptive map, the most medially situated tissues are absent, so that the neural groove has no floor, and, except when the cut is very close to the node, there is no chorda. The piece is, therefore, split for a greater or lesser extent up the centre, and assumes a V-shaped or U-shaped form (Figs. 25 b, 26).[1] On each side of the V, some neural tissue, representing the wall of the neural tube, may be present together with somitic mesoderm; these have presumably moved backwards nearly normally from their original more anterior positions. In some cases the thickened edges occupied by neural tissue project forward from the main cut margin; a similar effect was also noticed by Waterman (1936) and Rudnick (1938a), who repeated Waddington's experiments at a later date. This is probably due to the retraction of the cut edge in the regions of each side of the stiffened limbs of the V. Spratt (1947c) found the same type of V formation, but he had suggested, on not very obvious grounds, that the motive force for the backward movement is derived entirely from the elongation of the region just in front of the node, and since this is absent in the posterior segments, he could not allow that the appearance of the V is due to an autonomous movement; he suggests that it is wholly caused by a tearing apart consequent on the retraction. A similar suggestion was made by Jacobson (1938b), who denies the reality of any posterior movement during the regression of the streak. It may be admitted

---

[1] This splitting along the mid-line does not occur in the posterior halves of blastoderms sectioned in pre-streak, or short-streak stages (Spratt, 1942), before the presumptive axial tissues have been moved towards the anterior by the forward movements involved in streak formation.

that retraction and consequent mechanical tension may play some part in causing the two sides of the split blastoderm to gape widely apart; but it is not at all clear why a mere passive retraction should be such as to split the posterior part only, and to do that always in the mid-line, unless there was an autonomous tendency for this line to give way.

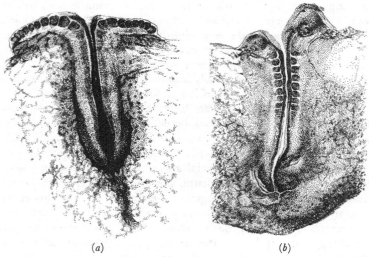

(a)   (b)

Fig. 26. *a* Posterior portion of a blastoderm sectioned at the 5-somite stage at a level just anterior to the node; the posterior part of the axis is split, and the two sides have been pulled apart by the retraction of the blastoderm; the somites extend further anteriorly than the neural tissue. *b* Posterior portion of blastoderm sectioned transversely through the node. (After Waddington.)

Spratt's main evidence against the above explanation of the phenomena is the fact that he did not observe a median split in many embryos from which he removed small fragments (of the presumptive 'eye-field', see p. 143) containing the node and the region just anterior to it. But, in the first place, inspection of the map of presumptive areas will show that the defect in this case only affected a very narrow strip of the median tissues; and, secondly, regenerative and regulative phenomena were very obvious. There is, therefore, no reason to expect any great development of the split; and it was pointed out as long ago as Waddington's paper (1932, p. 189) that the split is strongest when the level of section is some distance behind the primitive pit.

It is interesting to note that the neural tissue in these posterior isolates is joined up with the endoderm. There seems to be considerable affinity between these two tissues at this time, and they adhere preferentially to each other rather than to mesoderm.

It is of some importance to discover how far posterior to the node the cut can be made and chorda or neural tissue still develop. As regards chorda we have little information, although Waddington (1932) described a case in which it had developed when the cut was made very slightly posterior to the primitive pit (i.e. centre of the node) in an early head process stage; and it may be remarked that here there were two pieces of chorda, one on each side, which presumably implies that the presumptive area does actually extend in arc as it has been drawn in Fig. 10.

The cut can be much further back and still allow neural tissue to develop. It will be remembered that the chorio-allantoic workers could not obtain neural tissue when the node was excluded. Hunt (1932) expanded the necessary region to the 'node-field', which, in a special series of experiments, he found to extend no more than 0·2 mm. posterior to the primitive pit, that is, some 10 % of the length of the streak. In vitro, however, neural tissue develops on posterior segments when this distance is as much as 0·7 mm., that is, about a third of the streak length (Waddington, 1935). In later work, the chorio-allantoic school also began to have more success in this respect. Thus Dalton (1935) found some neural differentiation (although feeble) in pieces removed up to about 0·5 mm. posterior to the node, but claimed that this occurred only in presence of hypoblast, which was held to be a crucial factor. Rudnick (1938a, b), in investigations specifically aimed at a reconciliation of the in vitro and the chorio-allantoic results, found that hypoblast made no appreciable difference, but she was able to obtain neural differentiation in chorio-allantoic grafts only occasionally in strips which contained the second quarter (counting from the node backwards) of the streak; in these the anterior cut would be removed some 0·3–0·4 mm. from the pit. Moreover, she showed that the lateral walls of the neural tube, which develop on the sides of the V's in cultured fragments, fail to maintain themselves so as to differentiate into later neural tissue when transplanted to the chorio-allantois. Although this failure is probably the main explanation of the difference between the results of the two methods, it does not, of course, tell us anything about the process by which the tissue is originally formed, which was the question at issue. It provides a good example of the dangers, in experimental embryology, of investigating a process which occurs at a certain time by a method which does not allow one to assess the result until a much later stage after the performance of a large number of intermediate developmental steps; only too often these steps may turn out to be not as straightforward as one would wish and as is necessary if one is to argue directly from the far-removed result to the immediate situation one is hoping to study. Rudnick makes the just remark that her results 'emphasise the variability and selectivity of the chorio-allantoic environment and the deficiencies of the culture methods used'.

The same author also found the interesting fact that a transverse strip of blastoderm which, if it were cultured while attached to the rest of the posterior area, would undoubtedly form neural tissue, may fail to do so if it is isolated by two cuts, one say 0·4 mm. posterior to the pit and the other 0·9 mm. Such a strip undergoes profound morphological changes, which can perhaps be considered as attempts to perform the backward streaming in the mid-line which would be its presumptive fate. Actually, this involves a great elongation of the middle section, which becomes a thin contorted strand; but the presumptive neural tissue which it should contain does not develop as such. Whether it is the abnormality of the morphogenetic events which suppresses neural differentiation, or some other cause, cannot be stated.

In contrast to what happens in such relatively inhibitory environments as the chorio-allantois, parts of the primitive streak when grafted between the epi- and hypoblast of young embryos seem able to develop into their presumptive fate, and indeed perhaps into more than that. Thus Waddington (1932) described several cases in which grafts of the middle third of the streak (i.e. in which the anterior cut would be at least 0·7 mm. from the node) have developed neural tissue and even complete neural tubes. Probably many of these represent paragenesis of effectively isolated fragments. The ability of presumptive axial mesoderm to form neural tissue is well known in the Amphibia, both in cultures in salt solution, and still more in isolation sites in living animals. On the other hand, there is no doubt that in some of these intra-blastodermal grafts, the fate of the isolate is profoundly influenced

by its surroundings, particularly by the organisation centre of
the host embryo, and probably by that of any structure which
the graft succeeds in inducing (see p. 115). The development of
isolated fragments taken from in front of the node is discussed on
p. 143.

Little mention has yet been made of the most posterior part of
the streak. Its developmental capacities are not great in any

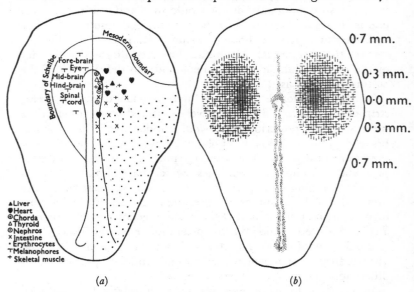

(a)   (b)

Fig. 27. *a* Map showing the regions of the definitive primitive streak blastoderm from which various tissues differentiate in chorio-allantoic grafts;
ectodermal tissues on left, mesodermal on right. (After Rudnick.) *b* Regions
of the head process blastoderm from which heart muscle differentiates in chorio-
allantoic grafts; the depth of the cross-hatching indicates the relative frequency
of heart differentiation. (After Rawles.)

environment. In tissue culture, and in intra-blastodermal grafts,
it usually forms blood (Murray, 1932; Waddington & Schmidt,
1933). Murray shows that this tissue may be formed from any
part of the posterior three-quarters of the streak.

It is clear from what has been said above that the technique of
culture has a powerful influence on the nature of the tissues
developing in an isolated fragment. But there is abundant
evidence that in some environments, such as the *in vitro* culture,

or in intra-blastodermal grafts, neural tissue can be formed from
parts of the streak which lie considerably to the posterior of the
node. The argument that the node is essential to neural differentiation therefore falls to the ground, and with it the further deduction—in any case an unjustified one—that the node is an
organiser. We shall see later that organising ability can be directly
tested in other ways.

Owing to the variable results of different methods, we can
compare the developmental capacities of different parts of the
blastoderm only if we confine our attention to data collected by
one particular technique. It is only by chorio-allantoic grafting
that a systematic exploration of the blastoderm has been under-
taken. Willier & Rawles (1935), Rawles (1936) and Rudnick
(1944) have brought together the data accumulated by the
numerous workers with this method, and expressed them in
a series of maps, showing the tissues which are formed by the
different areas (Figs. 27, 28).

It has already been pointed out that most of these grafts
contained cells from all three of the fundamental layers. We must,
therefore, be prepared to consider the possibility of inductive
reactions between them, and we shall see later that the axial
mesoderm in the chick, as in the Amphibia, is an organiser which
can influence the differentiation of both ectodermal and endodermal tissues in contact with it. Thus the map showing the
regions from which the mesodermal organs may be derived is
probably the most interesting.

To interpret this map, we should first compare it with the map
of presumptive areas, in order to determine whether the mesodermal tissues in isolates develop only in accordance with their
presumptive fate, or whether we are confronted with cases of
bedeutungsfremde differentiation. If the latter is found to be the
case, we have two apparently rather different, but perhaps really
related, types of explanation which our present knowledge suggests.
In the first place, we have seen that in the pre-streak stage, the
whole blastoderm is extremely labile; the situation in which the
mesoderm will be formed, that is to say, the location of the
organisation centre, is not fixed; and if such lability persists, in
some degree, until the streak stage, we must contemplate the
possibility that, in effect, new organisation centres arise within
the isolates. Secondly, we know that in the amphibian gastrula,

at a stage when the position of the mesoderm as a whole is firmly determined, isolated fragments of this tissue usually develop into a more diverse set of tissues than their presumptive fate would indicate. The flexibility is very far-reaching but differs in different groups. For instance, in both Urodela and Anura, pure presumptive mesoderm may develop into neural tissue, and while, in the former, chorda may be produced from the presumptive somite region, in the latter it is more nearly confined to its own presumptive area. Thus, in general, in the amphibian gastrula the areas from which neighbouring mesodermal tissues may develop in isolates, although centred on the presumptive areas, overlap quite largely.

The situation revealed in the chick by the chorio-allantoic grafting experiments is not easy to assess. It is convenient to begin by considering the results obtained in isolates from the head process stage, by which time the location of the main mass of presumptive mesoderm must certainly have been finalised. The maps of Willier & Rawles (1935), Rawles (1936) and Rudnick (1948) show that various tissues characteristic of the axial mesoderm (e.g. muscle, cartilage, nephros) can be found in isolates taken from some distance laterally to the streak, and the same is true, to a lesser extent, in the head-process stage mapped by Rudnick (Fig. 29). Unfortunately, we do not know exactly how far towards the sides their presumptive rudiments move before condensing again towards the mid-line during the formation of the body (p. 37). One cannot, therefore, be entirely certain that any *bedeutungsfremde* differentiation of mesoderm occurs at this stage; certainly the relatively specific localisation of the various 'potencies', insisted on by the chorio-allantoic workers, is evidence that the lability is restricted; nevertheless, the general impression is made that a given type of tissue can be obtained from an area larger than the exact presumptive fate of the rudiments would indicate. This opinion is shared by most of those who have carried out the experiments concerned. If it is adopted, one may conclude that the epigenetic situation in the chick mesoderm at this stage is very similar to that in the amphibian mesoderm towards the end of gastrulation; that is to say, there is a rough localisation of the various types of tissue, but the region from which a given type can be produced in an isolate is larger than its normal locus of origin. In the chick, heart muscle is, perhaps, the tissue which

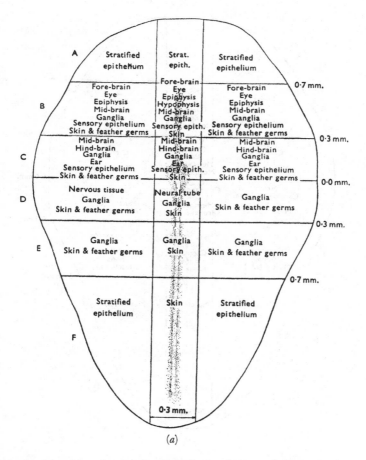

Fig. 28. *a, b, c.* Maps showing the tissues (respectively ectodermal, mesodermal and endodermal) differentiating in chorio-allantoic grafts from various regions of the head process stage blastoderm. (After Rawles.)

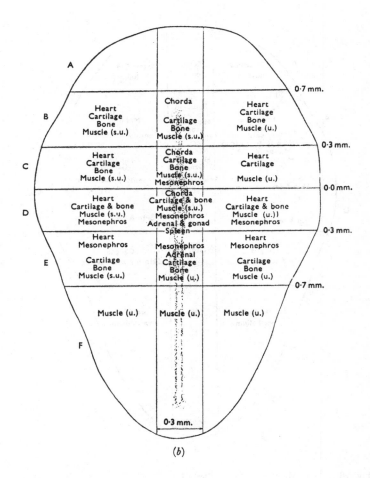

(b)

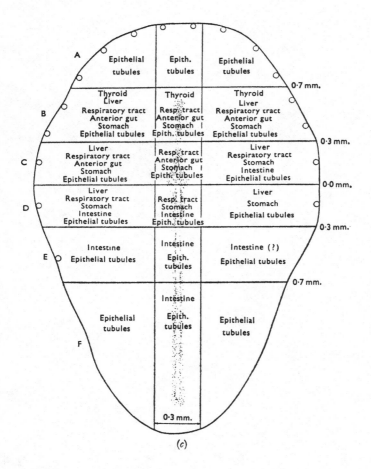

A	Epithelial tubules	Epith. tubules	Epithelial tubules	
B	Thyroid Liver Respiratory tract Anterior gut Stomach Epithelial tubules	Thyroid Resp. tract Anterior gut Stomach Epith. tubules	Thyroid Liver Respiratory tract Anterior gut Stomach Epithelial tubules	0·7 mm. 0·3 mm.
C	Liver Respiratory tract Anterior gut Stomach Epithelial tubules	Resp. tract Anterior gut Stomach Epith. tubules	Liver Respiratory tract Stomach Intestine Epithelial tubules	0·0 mm.
D	Liver Respiratory tract Stomach Intestine Epithelial tubules	Resp. tract Stomach Intestine Epith. tubules	Liver Stomach Epithelial tubules	0·3 mm.
E	Intestine Epithelial tubules	Intestine Epith. tubules	Intestine (?) Epithelial tubules	0·7 mm.
F	Epithelial tubules	Intestine Epith. tubules	Epithelial tubules	

0·3 mm.

(c)

extends most widely beyond its presumptive area[1] (Rawles, 1943). Compare Fig. 27 b.

In one respect, however, the situation seems to be somewhat the opposite of that which has just been described. The areas from which certain axial mesodermal tissues can be derived seem to be more limited in their posterior extent than would be expected from the presumptive maps. This may be true of chorda, and is almost certainly true of striated muscle, i.e. somite-mesoderm. We shall find that this is even more strikingly the case in the pre-head-process stage; it is probably to be attributed to the inhibitory effect of the conditions of grafting on the performance of the gastrulation movements in the streak.

Turning to the maps showing the origin of ectodermal and endodermal tissues from isolates in the head-process stage (Figs. 28 a, b, c) we seem, in regard to the latter, to be confronted with a situation similar to that described in the mesoderm. Although our knowledge of the presumptive map is not accurate enough to enable certain conclusions to be drawn, it may be surmised that endodermal derivatives such as liver, intestine, etc., can originate from regions outside the areas from which they are normally derived. We have available, in the first place, the same type of explanation which was suggested above, namely that localisation within the presumptive tissue is still under way, and the rudiments are not yet precisely delimited. It is also necessary to consider the possibility that these endodermal tissues owe their origin to inductive influences proceeding from the still labile mesoderm. It may be remembered that in Amphibia, Holtfreter (1938a, b) found that, on the whole, the endoderm of the gastrula showed less lability than the mesoderm; one gets the impression that in the chick the relative situation may be reversed.

In the ectoderm, it is clear that epidermal derivatives, such as feather germs, can originate over a wide area, which might suggest that they are independent of any stimulus from the axial mesoderm; in Amphibia, also, epidermal glands can develop in a similar way. The distribution of ganglia and melanophores is also easily comprehensible, in relation to the region which gives

---

[1] It is, perhaps, typical of the confused thinking and imprecise terminology of the 'chorio-allantoic' school that Rawles refers to these as 'heart-specific areas'. They are not, of course, specific for heart in the normal sense, namely, that they will not develop into anything else.

rise to neural tissue. The two former are known, in the Amphibia, to represent 'weak' grades of neural induction (cf. Gallera, 1947); it is, therefore, not surprising to find that the area from which they can be obtained is similar to that which gives rise to neural tissue, but extends out beyond it. The most important question which requires discussion is whether the area from which neural tissue develops is greater than that in which it is reasonable to suppose that inducing mesoderm is present. If it were so, we should have to postulate a tendency for self-differentiation of neural tissue, a situation which has been shown not to occur in Amphibia. In point of fact, the area is, for the anterior part of the nervous system, almost exactly the same as the presumptive area of origin. Since by this time the invagination of mesoderm must be more or less complete in the anterior region, we may conclude that all the anterior neural plate which in normal development becomes underlain by mesoderm will already have reached this condition by the time the isolations were made, and the development of it in the isolates may, therefore, be attributed to induction. The only part of the brain to which this argument does not apply is the fore-brain; the situation in this region of the embryo is discussed elsewhere (p. 140).

In the posterior region, conditions are quite different. The area from which neural tissue develops does not extend as far posteriorly as the presumptive map would lead one to expect. The problem before us is not whether there is evidence of a capacity for self-differentiation, but rather to find the reason for the failure of a differentiation which we might, at first sight, expect. A similar failure to achieve all that might be expected is even more striking when one examines the maps provided by Rudnick (1944) for the primitive-streak stage. At this stage, the failure occurs in the axial mesoderm as well as in the posterior part of the neural system. This might seem highly surprising. It is a very general observation in epigenetics that at earlier stages fragments of the egg are capable of a wider range of differentiation than they actually perform during normal development; or, at least, if this is not true of the egg as a whole, it is found that there is some part of it (such as the grey crescent-organiser region of the Amphibia) of which it is true; and we should expect the axial mesoderm to be the region in question in the case of the birds. Moreover, we have seen that at a stage still earlier than the streak, isolated parts (cultured by methods other than chorio-allantoic grafting, e.g. in Lutz's experiments) can perform much more than they normally do. Further, we have definite evidence, summarised above (p. 80) that, in the primitive-streak stage itself, isolates from the middle of the streak may *in vitro* develop into a wider range of tissues than they do on the chorio-allantois. It is difficult, therefore, to avoid the conclusion that the restriction of the areas of origin mapped by the chorio-allantoic workers at the primitive-streak stage to something less than the presumptive areas, is evidence not of a fundamental limitation of epigenetic capacity, but of an inhibitory influence of the conditions to which the graft is subjected. The regions which are most affected by this restriction are those which would normally undergo the rather complex morphogenetic movements of gastrulation, or those in which particular types of differentiation would normally be induced during the process of invagination. The inhibition, therefore, probably in the main affects these movements.

The evidence, therefore, seems to suggest that there is a series of phases in the development of isolated fragments of the blastoderm. In the earliest stages, before or shortly after the appearance of the streak, the whole germ is still in a very labile condition. Fragments isolated on the chorio-allantois are in general rather inhibited in differentiation, but the evidence of Lutz and Twiesselmann's isolations *in ovo* and of Waddington's endoderm rotations *in vitro* show that any part of the epiblast is still capable of developing into the streak and giving rise to mesoderm which exerts the inductive influences which will be discussed in the next chapter. As development proceeds, this lability becomes progressively restricted. First, the streak appears, and the site of mesoderm formation becomes fixed. It is probable that by the medium streak stage, no part of the epiblast outside the streak is any longer capable of giving rise to axial mesoderm as a response to mere isolation, although it may still be able to do so if stimulated by an inductive graft. Regional localisation within the mesoderm is, however, still quite fluid, so that isolated parts of the presumptive mesoderm may differentiate into tissues other than their normal fate. The expression of this property is, on the chorio-allantois, somewhat obscured by the inhibiting influence which that situation exerts on tissues which should undergo the invagination movements, but the tendency is exhibited by some mesodermal tissues such as heart muscle, and is also shown by the endoderm. Very diverse types of tissue are also produced by the region of Hensen's node, probably mainly by cells which have already invaginated, and as a response to inductive influences proceeding from these; it may well be that at this stage presumptive mesoderm may give rise to neural tissue, as it does in Amphibia. By the time the head process has grown to its full length, the localisation of the different types of mesoderm has proceeded further, each tissue now being developed only in isolates taken from the normal area of origin or from an area only slightly larger than this. There appears to be no necessity to postulate a capacity for autonomous differentiation in the neural tissue, except perhaps for the fore-brain.

Relations between mesoderm and endoderm

One particular inductive reaction between mesoderm and endoderm has been emphasised by the chorio-allantoic workers. It is found that liver (Willier & Rawles, 1931 a) and thyroid (Rudnick, 1932) are strikingly and constantly associated with heart in the mass of tissue into which the graft develops. This has been attributed to an inductive influence of the latter, but as Rudnick more cautiously points out (1935), it is not quite clear that the association is not merely the result of the spatial contiguity of the presumptive areas. Wolff (1933c), as a result of his studies of omphalocephalic embryos, showed that in these a liver rudiment may develop on each side. He summarises the situation in the statement that the liver 'se constitue au voisinage des veines qui se rendent au sinus veineux, à partir de deux ébauches situées à l'extrémité des ailes antéro-latérales de l'ébauche du tube digestif'. He does not discuss the possibility that the liver is induced by a paired mesodermal rudiment, such as that of the heart or veins, and seems to imply that it is self-differentiating. It is true that in some of his embryos in which the liver is double, the heart may be single and considerably removed from it, so that it seems improbable that the original suggestion of an induction of the former by the latter can be sustained; but it remains possible that some other region of the mesoderm is responsible.

The relation between mesodermal and endodermal organs has also been studied, in a more general way, by the separate cultivation of these two layers. We have seen that most authors (other than Dalton, 1935) find that it makes little difference to the development of a graft on the chorio-allantois whether hypoblast is left attached to it or not. All authors also agree that the isolated endoderm is unable to differentiate in grafts. The same is true *in vitro* (Waddington, 1932). On the other hand, epiblasts from which endoderm has been removed can, from at least a middle primitive-streak stage onwards, develop comparatively normally

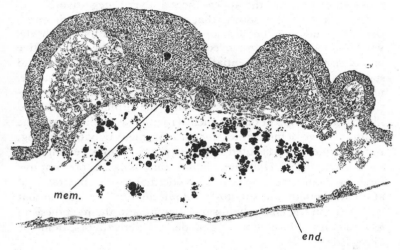

Fig. 29. Membrane (*mem.*) developed on the lower side of the mesoderm of a blastoderm cultivated *in vitro* after removal of the hypoblast in the streak stage. Lateral endoderm has also grown in centripetally on the surface of the clot (*end.*). (After Waddington.)

in culture, although the gross morphology may be somewhat distorted by the tendency to form thin-walled epithelial blisters.

In describing the development of isolated epiblasts, Waddington (1932) drew attention to indications which suggested a regeneration of the endoderm. In culture, the endoderm of the germ wall frequently grows inwards so as to cover the whole area below the epiblast; but this endoderm of extra-embryonic origin shows no sign of forming a foregut or being able in any way to differentiate into the organs normally produced by the embryonic endoderm. However, one also finds that the lower side of the mesoderm, which is in contact with the clot or the fluid which bathes its surface, frequently becomes arranged into a definite membrane.

Moreover, folds appear on the lower surface which strongly resemble the folds which go to form the fore-gut. Thus the anterior end of the embryo comes to be provided with a fore-gut-like space lined with a membrane resembling the endoderm. The epiblasts do not continue developing in culture long enough for the appearance of any well differentiated tissues, which would make it possible to judge whether the endoderm had been functionally replaced or not; but the appearances are very suggestive.

Hunt (1937 a) has shown that on the chorio-allantois endodermal tissues can be developed from grafts from which the whole endoderm was removed. This may well be a regulative or *bedeutungsfremde* differentiation of the mesoderm of the kind suggested above. Hunt, however, favours another alternative, namely, that the so-called presumptive mesoderm normally contributes cells to the endoderm during the primitive-streak and slightly later stages of development (1937 b); as we have seen, he claims to have demonstrated such a movement by vital staining, but his evidence is disputed by Peter (p. 17). It may be remarked that in urodeles, Holtfreter found that isolated presumptive mesoderm often developed into ectodermal tissues, but rarely if at all into endodermal (1938 b); but it is not at all impossible that conditions in the chick should be different, and the hypothesis of *bedeutungsfremde* formation of endodermal derivatives cannot be disregarded.

CHAPTER VI

THE ORGANISATION CENTRE AT THE PRIMITIVE-STREAK STAGE

Transplantation experiments

Spemann's discovery in 1918 of the organisation centre in the Amphibia made it clear that, at least in that group of vertebrates, interactions between parts of the embryo play a major role in epigenetics. In the next decade several authors felt the temptation to apply similar concepts to the development of the chick, and a variety of suggestions were forthcoming from various sides. Gräper (1929 b) argued that endoderm formation should have an organising function, while Wetzel (1929), Hunt (1929) and Willier & Rawles (1931 b) urged that the part of the organiser is played by Hensen's node. Most of these authors were very conscious of the fact that they had no technical means for carrying out the crucial experiment necessary to prove the existence of an organiser. It had not yet been possible directly to demonstrate the effect exerted by one part of the embryo on other parts in its neighbourhood, since this demands a transplantation of the region to be tested into a new location within an embryo still able to react to it. Suggestions as to organiser action had to be based on the evidence of defect and injury experiments, which are notoriously ambiguous in implication, and on morphological homologies.

It was to overcome these technical difficulties that the technique of cultivating the blastoderm *in vitro* was introduced by Waddington (1930). The main difficulties in carrying out transplantation experiments on avian blastoderms are caused by the albumen and vitelline membrane which cover it and the fluid yolk on which it lies. When the blastoderm is cleared of both of these, and transferred to a transparent plasma clot, the embryo is very much more accessible.[1] Even in that situation, however, it is not so easily operated on as are amphibian embryos. In particular, cut edges of tissue do not heal together so well, and it is, therefore, somewhat difficult to ensure a good 'take' if a fragment of tissue

---

[1] Organiser grafts have been made *in ovo* by Cairns (1937) and successful inductions obtained.

is inserted into a hole cut in the blastoderm surface. In recent years, Abercrombie has developed sufficient skill to obtain considerable success with such operations, but in most of the earlier and more fundamental work the grafts were made in another way, the grafted fragment being inserted into a pocket between the epiblast and hypoblast. In preparing the pocket, the line of approach is usually through the hypoblast, so that the epiblast remains quite uninjured.

This technique has the advantage that it facilitates the recognition of the grafted tissues and the discrimination of them from those of the host. In the Amphibia, when it is desired that the host and graft tissues shall be separately recognisable, it is usual either to employ vital staining or to make transplantations between different species whose tissues are in some way dissimilar. Neither of these methods has yet been successfully employed in the chick. The usual vital stains are too fugitive in that form and attempts to use more permanent forms of 'staining'[1] have so far proved ineffective. Moreover, the tissues of early avian embryos of different species are very similar, and cannot be separately recognised. Thus, for the recognition of the graft, one is thrown back on to morphological criteria. Very often, as will be seen from the figures, a lump of tissue, extraneous to the normal anatomy of the host is found in the region into which the graft was made. Much of this lump is likely to be quite disconnected with the host epiblast; and such parts can usually be considered to be derived from the graft. There may also be parts which are directly continuous with the epiblast, and since the latter was not injured in the operation, this continuity is not likely to be due to a secondary healing together of host and graft tissues; such portions can then be interpreted as the results of the reaction of the epiblast to inductive influences proceeding from the graft. Induced neural tissue can usually be recognised with considerable certainty in this way, provided the embryo has not developed to such a stage that the neural tube has lost its connection with the epidermis. It is, however, much more difficult to discriminate between graft and induced mesoderm. In fact, in many cases this is impossible and the best we can do is to state that it looks as

---

[1] E.g. by allowing the cells to pick up colloidal iron which could later be made visible by cytochemical methods (Waddington, 1937, unpublished).

though both are present although the boundaries between them are uncertain.[1]

Although the inapplicability of the normal methods of differentiating between host and graft tissues in the chick renders the interpretation of transplantation experiments somewhat less clearcut than it is in the Amphibia, the importance of this point should not be exaggerated. The morphological criteria are usually in practice rather more convincing than they sound in the abstract, and none of the major points which will be discussed in this chapter depends on specimens in which there remains much doubt as to the interpretation.

The reality of the induction phenomenon

In order to circumvent as far as possible the uncertainties of recognising graft and host tissues, Waddington (1930, 1932) demonstrated the occurrence of embryonic induction in the primitive streak stage by a method in which both inductor and induced retained their anatomical continuity with the whole blastoderm to which they belonged. The hypoblasts were removed from each of two blastoderms, and the epiblasts then placed with their mesodermal faces together, in such a way that the two primitive streaks did not lie in contact throughout their whole length. On cultivating the double-epiblast *in vitro*, it was found that the lower one, which lay with its ectoderm side against the surface of the plasma clot, fairly soon suffered necrosis and disintegration. However, some differentiation took place even in it, while in the upper epiblast it proceeded nearly normally.

In such cultures, there may occur as many as four embryonic axes, each comprising a neural plate with associated mesoderm. Two of these are derived from the two original primitive streaks; the other two, one or other of which may be absent in particular cases, lie immediately in contact with the former, but in the other epiblast; they must be the product of induction. All four neural plates are in full anatomical continuity with one or other of the epiblasts, and there seems no possibility of suggesting that an

[1] Abercombie & Causey (1950) have recently introduced a new technique. They inject a small quantity of radioactive phosphorus (^{32}P) into the albumen, and show that grafts made with parts of the embryo developing in such an egg can later be detected and to some extent distinguished from host tissues by auto-radiography.

apparently induced plate is really part of the presumptive neural tissue of the other epiblast, which has broken loose from it and joined up with the epiblast in which it is now found. The proof of induction is, therefore, as complete as could be wished (Fig. 30).

More recent work, in the Amphibia, has shown that necrotic tissue often acquires a power of induction which it did not previously possess, and it might be asked whether the necrosis of the lower blastoderm plays an important role in these experiments. The answer would seem to be in the negative. In the first place, the induced axis in the upper epiblast is always in contact with the inducing axis in the lower epiblast and not necessarily with

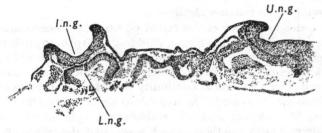

Fig. 30. Induced neural groove (*I.n.g.*) lying above the neural groove (*L.n.g.*) of the lower of two epiblasts cultivated after removal of the hypoblasts and with the mesoderm faces together. The axis of the upper epiblast is at *U.n.g.*; the lower epiblast is extensively degenerated. (After Waddington.)

the most necrotic region. Secondly, induction sometimes takes place downwards, from the more healthy upper axis into the dying lower ectoderm. And, finally, we shall see many other cases of induction by grafts of the primitive streak in which no abnormal necrosis is to be found.

Organiser grafts

A considerable number of grafts of pieces of primitive streak and other regions of the blastoderm have been made by Waddington and his pupils, Abercrombie, Waterman, Taylor and Schmidt, and a few by Woodside in America. In some cases, the result is merely the differentiation of the graft itself without any marked reaction by the host. This occurs particularly when the graft lies beneath a layer of host mesoderm, which seems to isolate it from the host ectoderm. It also occurred frequently, for some not very clearly defined reason, in the grafts made by Schmidt (cf. Waddington & Schmidt, 1933), and the same seems to have been true in Woodside's series (1937). Possibly it was the use of larger fragments of tissue as grafts which was to blame for the comparative lack of success by these workers.

There are, however, very many cases in which grafts have exerted a clear-cut inductive effect. A series of these are illustrated in Figs. 31–33. These specimens demonstrate, first, that induction can be performed by pieces of the anterior part of the

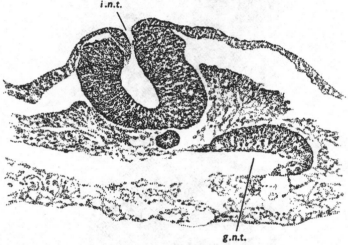

Fig. 31. A graft of the anterior third of a primitive streak has yielded a small piece of neural tissue (*g.n.t.*) and perhaps some mesoderm, and has induced a neural tube and notochord (*i.n.t.*). (After Waddington.)

streak. The presence of Hensen's node is not necessary (Fig. 32 *a*, *b*). On the other hand, inductions have never been obtained from the posterior third of the fully grown streak. The boundary between the inducing and the non-inducing region appears to be about at the mid-point of the streak. This is near the limit at which we have seen (p. 82) that neural tissue can develop in posterior fragments isolated *in vitro*; and it is indeed usual to find that some neural tissue is produced by grafts which succeed in induction. However, a few cases are known in which this is not the case (Fig. 33). The boundary of the presumptive axial mesoderm probably also lies in much the same region, and it seems fairly safe to attribute the induction primarily to this tissue rather than to the presumptive neural tissue. This, of course, brings the chick strictly in line with the Amphibia. As in that group we must be prepared

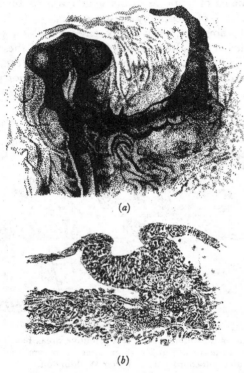

(a)

(b)

Fig. 32. A graft of the anterior half of the streak (excluding the node) from a head-process duck blastoderm into a long streak chick blastoderm has induced a structure joined on to the host's axis, as shown in a. Sections show that, near the host axis, this consists of an induced neural groove b which extends into a finger-like process containing streak-like tissue. (After Waddington and Schmidt.)

for the possibility that, in grafts, axial mesoderm with inducing powers may sometimes be formed (by regulation) from neighbouring non-presumptive axial mesoderm, so that the boundaries of the organisation centre may be slightly different from those of the presumptive tissue; but in the chick our knowledge of the presumptive map is not accurate enough for this to be tested.

In the chick induction can be performed by the immediate product of the primitive streak, namely, the head process; and also by the sinus rhomboidalis, which after all is little other than a persistent streak remaining in the posterior end of the later

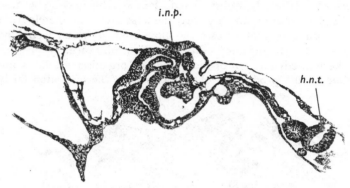

Fig. 33. Induced neural plate (*i.n.p.*) accompanied by a fore-gut, formed in response to a graft of the posterior two-thirds of the streak. The graft has formed no neural tissue. The host neural tube is *h.n.t.* (After Waddington.)

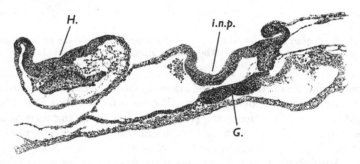

Fig. 34. Induced neural plate (*i.n.p.*) formed in response to a graft of the embryonic axis from just posteror to the last somite of a 14-somite embryo. *G.* graft, *H.* host axis. (After Waddington.)

embryo (Fig. 34). It can also be carried out by pure neural tissue, free of mesoderm, derived from early somite embryos (Fig. 35). This is the phenomenon of 'homoiogenetic induction' described by Mangold & Spemann (1927). Attempts to demonstrate an inducing power in the chick notochord (Waddington, 1933*a*) have so far been unsuccessful and in grafts of the head process and sinus rhomboidalis, the extent of the induced plate seems to be related far more to the extent of the graft neural tissue and somitic mesoderm than to that of the graft chorda. It seems likely, then, that there is little inducing capacity in the chick notochord, after it becomes definitely differentiated. Perhaps this may be because in the chick that organ very early detaches itself completely from other tissues, so that it loses the close contact which would be necessary for the diffusion of an inductive substance.

The inductive action of these various 'organiser grafts' is not limited to the production of neural tissue. There is often fairly

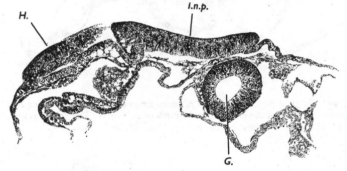

Fig. 35. A graft *G.* of neural tissue from the brain of an 8-somite embryo has induced a neural plate (*i.n.p.*). The anterior tip of the host neural plate is cut at *H.*

convincing anatomical evidence that the host has been stimulated to form axial mesoderm as well. We may indeed find that the epiblast overlying the graft gives every sign of having formed a primitive streak, in which the invagination of mesoderm is actively proceeding. This is again parallel, of course, to what is found in the Amphibia, in which an organiser graft frequently induces the formation of a blastopore and the invagination of host tissues, which then develop into mesoderm.

A similar induction of presumptive ectoderm to form mesoderm was demonstrated by Waddington & Taylor (1937) in a special investigation (Fig. 36a). The technique was slightly different, in that fragments of presumptive ectoderm were grafted into holes made in the primitive streak of a host blastoderm. In the chick, the conversion occurs only if the grafted fragment actually under-goes an invagination process; mesoderm is not formed if epiblast is grafted directly under the streak (see p. 115). The situation is different in the urodeles, where Mangold (1925) and Bytinski-Salz (1929) found that presumptive ectoderm, grafted into the mesoderm through a hole in the gastrula roof, could become mesoderm without ever having invaginated through a blastopore in the normal way. Waddington & Taylor attributed the failure

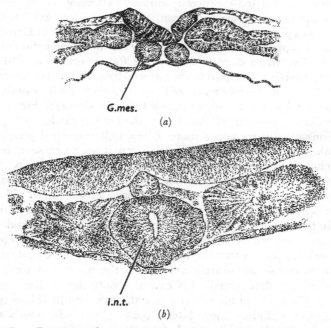

G.mes.

(a)

i.n.t.

(b)

Fig. 36. a. Excess mesoderm (G.mes.) formed from epiblast (non-presumptive mesoderm) grafted from a long streak stage blastoderm into the streak of another blastoderm of the same age. (After Waddington & Taylor.) b. An induced neural tube (i.n.t.) formed from area opaca ectoderm placed under the anterior end of a long primitive streak. (After Abercrombie.)

to observe a similar result in the chick to the fact that in that form the mesoderm originates as a mass of loosely connected cells, while the epiblast is a well-defined epithelium. They suppose that the formation of mesoderm demands that this epithelium should be broken down into its component cells, a process which may be less relevant in the Amphibia in which the archenteron roof preserves much of the epithelial structure of the blastocoel roof from which it is derived. Definite evidence was obtained that presumptive ectoderm under these circumstances may be converted into mesoderm, which fused to a greater or lesser extent with the host mesoderm. The result is similar to that recorded by Spemann & Geinitz (1927), who showed that in *Triton* fragments of presumptive ectoderm grafted into the blastopore lip so as to become invaginated are thereby induced to develop as mesoderm.

In the chick, as we have seen, the presumptive axial tissues undergo a great elongation connected with the retreat of Hensen's node. It appears that such tendencies for morphogenetic movements, like those directly responsible for invagination, can be transmitted from the graft to the induced tissue. Thus it is common to find that the posterior part of the induction forms an elongated projection lying above the surface of the blastoderm (cf. Fig. 32 a). In embryos fixed 24–36 hours after the graft is made, such projections usually contain a more or less well-developed primitive streak in their most posterior part, and it seems reasonable to conclude that they all originate as induced primitive streaks. The drawing out of the projection can then be attributed to the backward stretching which is a normal characteristic of the streak's development. It is worth pointing out that such projections may be induced by grafts which do not include Hensen's node, a circumstance which would hardly be expected if one accepted Spratt's suggestion that the tissues which build up the elongating embryonic axis lie entirely anterior to the node; but we have seen (p. 27) that actually his evidence only shows that this is true of the tissues of the mid-line and that we are still at liberty to believe that slightly more lateral parts of the axis, which also share in the elongation, are formed from presumptive material which lies behind the node in each side of the streak.

The induction of morphogenetic movements was clearly revealed in some experiments of Abercrombie (1937) in which fragments of epiblast were placed between the primitive streak and the hypoblast of a host blastoderm. In these circumstances the grafts developed, not into induced mesoderm, but into neural tissue (Fig. 36 b), even when they were derived from the area of presumptive mesoderm. This neural tissue showed signs of having been affected by two types of morphogenetic movement. One was an elongation in consequence of which the induction might become as much as three times as long as the graft had been at the time of operation; and naturally at the same time it would usually become narrower. The elongation did not carry the anterior end of the graft forwards; nor was there any evidence that the narrowing of the graft was due to necrosis of the edges. Further, the graft was never shifted backwards as a whole, as might have been expected if it had merely been passively affected by the host movements. Observation during the cultivation of the embryo showed that usually the elongation was brought about by the posterior edge of the graft moving backwards in company with the retreating node of the host. There seems no doubt in fact that we are dealing with the induction of an active morphogenetic movement of the kind proper to the region immediately in front of the node.

The second movement which affected these grafts was one which led to what Abercrombie called a 'centring'; that is to say, the induced neural tissue showed a strong tendency to lie perfectly symmetrically under the mid-line, even when it was known that the graft had not been perfectly centred at the time of operation. Some share in this effect can probably be attributed to the influence of the invagination movements of the host, but the centring of the posterior part of the induction, which was found during the posterior elongation, was usually more complete than that of the anterior part, and it is therefore likely that the phenomenon is partly to be explained by the hypothesis that it is in the exact mid-line that the tendency to induce the backward stretching is strongest.

It is worth noting that Abercrombie also described cases in which neural induction had occurred without being accompanied by any verifiable elongation. The morphogenetic movement is therefore not a necessary part of the induction process.

Organiser grafts may also affect the endoderm. In several of the examples shown in the figures, a well-formed fore-gut was found in connection with the induction. Although in some cases endoderm was included in the grafted fragment, these induced fore-guts were usually so thoroughly continuous with the host endoderm that there is no doubt that some inductive influence had been at work on the host tissues, and it is probable that in many cases the induced fore-gut was wholly formed as a response to induction, there being no important contribution from self-differentiating graft tissues. This must certainly be the case when the graft contained no endoderm, unless we adopt Hunt's suggestion that the axial endodermal structures are formed from material invaginated through the streak.

The induction of the fore-gut by the streak is best demonstrated in those blastoderms in which the axes of the epi- and hypoblasts have been mutually rotated (see p. 52) and in which the embryo has developed in the epiblastal direction. In these the gut is entirely formed by non-presumptive endoderm as a response to an inductive stimulus proceeding from the epiblast.

It seems probable that the inducing stimulus which acts on the endoderm proceeds mainly from the mesoderm. Waddington & Schmidt (1933, p. 549) pointed out that at that time no case had

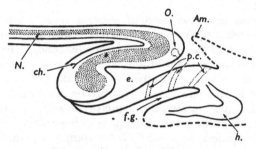

Fig. 37. Structure of the 'omphalocephalic monster'. The anterior end of the neural tube (N, dotted), underlain by the chorda (ch.), is bent down into an ectodermal pocket (e.); the foregut (f.g.) extends anteriorly, with the heart (h.) and pharyngeal clefts (p.c.) associated with it; otocyst at (O.) and the amniotic fold at (Am.). (After Wolff.)

occurred in which a fore-gut had been induced in the absence of neural tissue, either grafted or induced. But this certainly seems to occur in the 'omphalocephalic monsters' studied by Wolff (1933a,b, 1936). In these the head is bent down so as to form a hernia into the gut, while anterior to the head the endoderm develops a fold which pushes upwards towards the dorsal surface. This fold can be seen, from the character of its later differentiation, to represent the fore-gut. It appears almost certain, from the early specimens described by Wolff (1933b), that the abnormality occurs before or very shortly after the period in which the fore-gut-induction must take place, and it would therefore appear that this induction cannot be attributed to the neural tube, which in these embryos never comes into close contact with the endoderm developing as fore-gut. However, Wolff shows that the heart vesicles also develop anteriorly to the head, in the immediate neighbourhood of the fore-gut. It therefore seems probable that it is to the mesoderm that the fore-gut induction is due, unless of course the differentiation in this case is autonomous and owes nothing to any inductive stimulus.

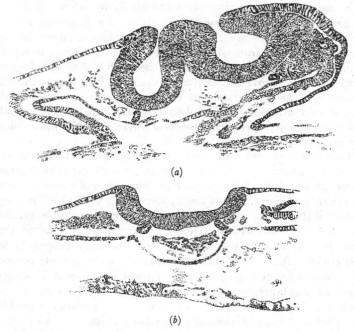

(a)

(b)

Fig. 38. a, b Two sections through an embryo in which an induced axis (on right) has joined up laterally with the host axis.

Fairly soon after its formation, the fore-gut begins to form lateral diverticula, with thickened walls, which are the forerunners of the pharyngeal pockets. Abercrombie (1937) and Waddington (1937a) described a number of cases in which mesenchyme derived from primitive streak grafts lay to one side of the head mesenchyme of a host embryo. In these specimens, fore-gut diverticula were never present at the edge of the host head mesenchyme, which would be their normal place of occurrence. Instead they were found at the edge of the whole mass of mesenchyme, in the region where it adjoined the lateral plate mesoderm (cf. Fig. 38). They therefore suggested that the thickening of the fore-gut epithelium in the diverticula is induced as a response to some stimulus dependent on this junction between head mesenchyme and lateral plate.

The analysis of induction: competence and self-differentiation

The events which follow the grafting of a fragment of the organisation centre are complex. They include changes in the histological types into which the neighbouring tissues develop, and the assumption of a greater or less degree of morphological organisation. To give an adequate account of them requires much more than the bare statement that an induction has occurred. We should attempt to formulate hypotheses as to the causal processes which underlie all aspects of the final results with which we are confronted.

One of the elements in these causal processes is the tissue on which the inductor acts. It is, of course, not the case that a piece of the primitive streak can induce the formation of a neural plate from any and every type of embryonic tissue; the induction only occurs if the reacting tissue has some appropriate property. In the period before serious attempts were made to analyse the induction process it was usual to refer to such properties as 'potencies'. A region of embryonic tissue was then said to possess the potency for becoming some later tissue or organ A, if, under some circumstance or other, it did become A. The weakness of such a concept is that it is purely descriptive; it does not enable us to relate the act of becoming A to any particular event at any particular time. In the induction process, however, we are confronted with a definite causal interaction between inductor and induced. We may hope to define precisely the period during which the interaction occurs; and we must discuss it in terms of concepts which characterise the various reactants at the actual time of their reaction.

A concept of this kind was first precisely defined in connection with the induction process in the chick. Waddington (1932) proposed that the period during which a given tissue is capable of reacting with a certain inductive stimulus be called the 'period of competence' (the nature of the stimulus, and of the developmental result of the reaction, being specified if necessary). During this period, the reacting tissue can be said to be competent to react with such and such an organiser so as to develop towards such and such an end; or we may say that it possesses the competence for this end-result. It was pointed out that usually a given tissue possesses several competences at one and the same time, and the hypothesis was advanced that the state of competence is one in which a certain number of alternative paths of development are open to the tissue; but it is important to remember that the concept of competence loses all precision unless the period of competence is limited to that period during which the decision between the various alternative paths of development is actually being taken. Further discussion of the concept will be found in Waddington (1935, 1940).

Competence for the formation of neural tissue is present throughout the whole epiblast of the primitive streak stage, not only in the presumptive neural plate, and the ectoderm of the area pellucida, but also in the presumptive axial mesoderm[1] (Abercrombie, 1937) and the ectoderm of the area opaca (Waddington, 1934a; Abercrombie, 1937). The last of these has never been underlain by the embryonic endoderm, which we have seen (p. 55) to be capable of exerting an inductive influence on the morphogenetic movements which lead to the formation of the primitive streak; thus the competence for neural tissue formation cannot be dependent on any previous influence of this earlier 'organiser'.

It is probable that the whole epiblast at this stage also has a competence for the formation of axial mesoderm, since all the inductions made in various parts of the blastoderm have mesoderm associated with them, and some of this is likely to be derived from the host. However, the difficulty of distinguishing between host and graft mesoderm makes it impossible to state this conclusion with perfect conviction. The only absolutely definite proof of the existence of a mesodermal competence in the epiblast is that given by Waddington & Taylor (1937) for the fragments derived from the edge of the area pellucida or from the mesoderm-free region just anterior to Hensen's node.

Still less is known about endodermal competences. We have seen that Hunt (1937b) supposes that tissue, otherwise regarded

---

[1] It may be absent in the presumptive lateral mesoderm of the posterior part of the primitive streak, although it is present just posterior to the end of the streak (cf. p. 121).

as presumptive mesoderm, normally contributes to the embryonic endoderm, and he has good evidence that it may do so in chorio-allantoic grafts; and we have mentioned (p. 91) the 'false endoderms' formed from mesoderm in endoderm-less epiblasts in culture. It seems, therefore, that the mesoderm has competence for endodermal differentiation. Within the endoderm itself it is clear, from the induced fore-guts mentioned above (p. 104), that competence for gut-formation is spread throughout the endoderm of the area pellucida. Organiser grafts into the area opaca have never induced the formation of a gut from the local endoderm, which is perhaps not surprising in view of the great histological difference between it and the embryonic endoderm.

Some attention has been paid to the temporal as well as to the spatial aspects of competence. In the first place, it may be noticed that an induced neural plate seems always to be at the same stage of differentiation as the graft neural plate, even when there is a fairly considerable difference in the age of the two tissues (cf. Waddington & Schmidt, 1933, p. 525; Abercrombie, 1937, p. 308). This may entail the appearance of an induced neural plate in the vicinity of part of the host axis which has not passed the primitive streak stage (cf. Waddington, 1934, p. 213). Such a phenomenon demonstrates that the inductive process can, in this region, take place earlier than it normally does, and thus that the beginning of the period of competence may be earlier than the time at which induction normally occurs. The contemporaneity of the induced and inducing plates would then be a simple consequence of the fact that they are both simultaneously induced by the same axial mesoderm (cf. Abercrombie, *loc. cit.*).

Woodside (1937) has recorded an investigation into the temporal changes in competence. He claims that the most pronounced response to grafts of the primitive streak is obtained after transplantation to the short broad primitive streak, and that the ability of the epiblast to form neural tissue decreases with increasing age. He does not seem to have realised, however, that the earlier the graft is made the longer it has to act, so that the extent of the response cannot be taken as a direct indication of the condition of the responding tissue, since the stimuli are not constant. Moreover, the inductions figured by him are in general rather poor ones, presumably owing to technical inadequacies. It would seem very unsafe to draw conclusions from essentially quantitative comparisons between members of such an unsatisfactory series. This applies particularly to his statement that in the head process stage neural tissue can be induced in the anterior ectoderm, while the more posterior regions form only ectodermal thickenings. Since at this time it is possible that invagination and induction are still proceeding at the streak in the posterior region of the blastoderm, it would be very unexpected to find that the competence had already disappeared there. We should need much more satisfactory evidence than that offered by Woodside to reject the usual law of anterior-posterior development in this connection.

It may also be pointed out that there is a technical difficulty in such experiments which Woodside does not seem to have appreciated. In the late streak and head process stages, most of the epiblast is underlain by a closely adherent layer of mesoderm. A graft will have no proper chance to exert an inducing stimulus unless this mesoderm can be cleared away and prevented from insulating it from the ectoderm. It is extremely difficult to do this completely, or even to judge to how great an extent the inhibitory influence of the mesoderm has been circumvented. This difficulty is, in my opinion, so great as to render it impracticable to attempt a precise study of the waning of the neural competence in the chick epiblast. Woodside states that it has disappeared at the head fold stage; this is probably correct, if one takes it to mean that a region loses competence at about the same time as it actually forms recognisable neural plate, but the proof of it is still inadequate.

Closely allied to the question of competence is that of self-differentiation. One must inquire, that is to say, whether the tissues of the blastoderm can resolve the choice between the various alternative paths of development open to them, and enter on one of them, without the aid of an external inducing stimulus.

As regards the ectoderm, the question is not easy to answer unequivocally. We have seen that both Waddington (1932) and Spratt (1947c) found that fore-brain will develop from anterior parts of the blastoderm, isolated at a time when they would contain little or no mesoderm. It remains, perhaps, uncertain that absolutely no mesoderm of any kind is present; but at its face value, the evidence would seem to indicate that at least the pre-chordal region of the brain develops independently of any inducing stimulus derived from the mesoderm. It remains possible, however, that some stimulus emanating from the endoderm plays a part in bringing about this result. It must be remembered also that in the Amphibia the pre-chordal neural tissue (the 'acrencephalic region' of Dalcq, 1947, or the 'archencephalic region' of Lehmann, 1945) shows a certain autonomy as against the remainder of the neural system and appears to be induced by the pre-chordal plate rather than by the archenteron roof *sensu stricto*.

Even if we suspect a certain ability for self-differentiation of this region in the chick, we have no right to extend this to cover the whole presumptive neural system. In fact, we have little positive evidence which would suggest any capacity of the rest of the presumptive neural tissue to develop into its neural fate in the absence of the appropriate inducing stimulus. The difficulty of attaining any complete separation between this tissue and the axial mesoderm makes a precise analysis of the situation impossible, but nothing speaks against the assumption that the situation is exactly similar to that in the Amphibia.

The same seems to be true of the self-differentiating capacities of the mesoderm, which have already been discussed on p. 85.

The analysis of induction: evocation and individuation

The suggestion that induction can be analysed into two aspects, evocation and individuation, has been discussed in a number of places already (cf. Waddington, 1941; Needham, 1942). In its relation to the chick material, in connection with which the concepts were first advanced (Waddington & Schmidt, 1933), the distinction between them may be clarified if we remind ourselves again of the nature of the phenomena with which the process of induction confronts us. Figs. 39 and 40 show the result of two organiser grafts (Nos. 32–135 and 32–136). They were grafts of the two halves of a medium length chick primitive streak into a duck blastoderm of similar age. In No. 32–135, the anterior half of the streak has induced a structure which clearly bears a very considerable resemblance to a head; that is to say, the response to the inducing stimulus is not a mere amorphous mass of neural and other tissues, but is, to some extent at least, a morphologically organised unity, which can be recognised as similar to an organ which we already know. In No. 32–136 also, the posterior half has induced an organised structure, though in this case the graft and the tissues which it has acted on have become joined to the structures of the host axis. In some respects, as in the neural plate, a more or less normal, though over large, structure is formed by the combination. Other organs, such as the heart and fore-gut,

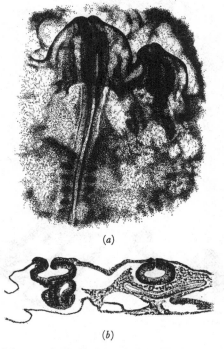

(a)

(b)

Fig. 39. Graft of the anterior half of the medium-length streak of a chick into a duck blastoderm of similar age: *a* shows the embryo in life after $26\frac{1}{2}$ hours, with an induced head to the right of the host head; *b* is a section through the middle of the induced axis (on left in figure) showing the induced neural tube lying above neural tissue and mesoderm derived from the graft, with the host axis to the right. (After Waddington & Schmidt.)

are duplicated; but it is remarkable that the heart of the induced axis is at the same transverse level as that of the host, and the like is true of the anterior end of the notochords, of the fore-gut openings and even of the intersomitic grooves. Since the graft consisted only of the posterior half of the streak, it is obvious that this parallelism cannot possibly be due merely to chance, but must be the result

112

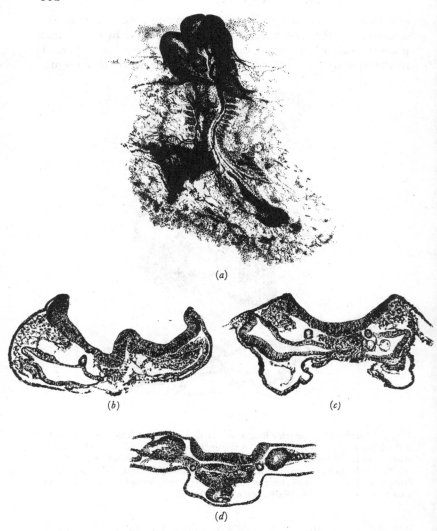

(a)

(b)   (c)

(d)

Fig. 40. Graft into a duck blastoderm of the posterior half of the same streak
as used in the specimen of Fig. 39: *a* shows the fixed specimen, after 45 hours
cultivation; an induced axis to the left of the host has united laterally with it.
b, *c*, *d* Three sections through the head, heart and somite regions of the
combined embryos (induced embryo to the right in the figures). (After
Waddington & Schmidt.)

of a mutual interaction between host, graft and induction, in which the host seems to have taken the lead.

These two inductions exhibit phenomena—of the formation of organised structures, of the assumption of a particular regional pattern characteristic of some definite level along the axis—which require discussion, and for which we therefore need an appropriate vocabulary. At the same time, it must be remembered that such phenomena are not invariably associated with induction; the induced tissues are frequently chaotic, having neither morphological organisation nor regional character. Moreover, as we shall see, there is evidence that sometimes the organisation of the induced tissue is directly transmitted to it from the inductor; but again, this is not a necessary property of an inductor, since we know of such which have no organisation to transmit.

We can therefore distinguish two aspects of induction. One, to which the name 'evocation' has been given (Needham, Waddington & Needham, 1934), is the mere process of inducing something or other—the selection of one or other of the developmental paths open to the competent tissue; it is the part of the induction process that can be brought about by a stimulus which has itself no organisation to impart, but simply pulls a trigger and leaves the rest to the reacting partner. In contrast to evocation, Waddington & Schmidt introduced the name of 'individuation' for the process of formation of a morphologically organised structure recognisable as the whole or part of a unit which can be regarded as having a certain individuality. Crudely speaking, in fact, individuation is the process of formation of an organ, and it necessarily suffers from such degree of vagueness as attaches to the concept of an organ: to pursue the definition further carries us into more or less philosophical realms (cf. Waddington, 1940a).

It was suggested that the whole complex of phenomena involved in induction can be adequately brought under these two heads of evocation and individuation. This involves the hypothesis that not only the induction of structure but also that of regional character have an essential similarity with the processes of unit-formation which we have taken as the paradigm of individuation. This point will be discussed in detail below.

Evocation

The chick embryo has not proved very suitable for the study of the nature of the evocating stimuli involved in induction. The proof that an evocation stimulus can be exerted by tissues of the primitive streak even after they have been killed by heat coagulation was provided (Waddington, 1933*b*, 1934*b*) in work carried on simultaneously with demonstrations of the same fact in the Amphibia (Bautzmann, Holtfreter, Spemann & Mangold, 1932). But a dead implant into the chick blastoderm is usually

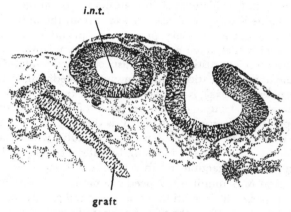

Fig. 41. Induced neural tube (*i.n.t.*) formed as a response to a graft of primitive streak coagulated by a short immersion in boiling water. (After Waddington.)

soon covered by a layer of mesenchymatous cells, which appear to insulate it from the host ectoderm, which in consequence rarely shows any reaction.

It is perhaps worthy of note that one of the inductions produced by a dead implant included an induced chorda as well as neural tissue; this is parallel to the phenomena found in Amphibia.

As might be expected from what we now know of the process of evocation, the chick primitive streak can produce inductions in embryos of very distantly related types: for instance, in rabbits (Waddington, 1934*c*) and in amphibians (Hatt, 1934). On the other hand, few attempts have been made to obtain evocation with foreign tissues grafted into the chick blastoderm. Orts Llorca (1948) has described some rather doubtful successes following injection of dried extracts of human testis. Abercrombie (1939) has shown that some of the polycyclic hydrocarbons which are known to be powerfully evocating substances in the Amphibia, are also active in bringing about neural differentiation in isolated fragments of chick epiblast.

Individuation and regional determination

There are three elements which take part in the differentiation of structures in the neighbourhood of an organiser graft: they are the graft itself, the formations which it induces and the host embryo. In the latter, it is the axis which exerts the strongest influence, so that when the graft is in a locality far removed from the centre of the blastoderm little effect of the host may be discernible. But in general, we have to take account of all three elements. Each of them, it will be found, has at the time of grafting a certain disposition to develop into particular structures and regions; but these dispositions, although varying in intensity, are all capable of some re-direction under the influence of the surrounding materials. The discussion of the results which have been observed hitherto has been given mainly by Waddington & Schmidt, who introduced the term 'individuation' in their paper of 1933, and by Abercrombie & Waddington (1937), who recorded a large series of experiments in which fragments of primitive streak were grafted under the streak of a second blastoderm in order to provide material for studying the interaction between an uninjured organisation centre and a neighbouring organiser fragment.

It will perhaps be convenient to consider the developing structure in terms of its three axes, the dorso-ventral, the medio-lateral, and the antero-posterior.

(a) *Dorso-ventral organisation.* In the experiments which have so far been performed, the dorso-ventral axis of the ectoderm has never been changed. This has been particularly well shown in Abercrombie's (1937) experiments, in which pieces of epiblast were placed below the streak and became induced there to form neural tissue; the orientation of this, which can be recognised by the position of the mitoses in the epithelium, was always the same as that of the original graft (Fig. 36 b). The presence or absence of hypoblast made no difference, so that it appears that the dorso-ventrality is fixed within the epiblast itself. The same retention of the original polarity also appears to be found when it was presumptive mesoderm which became induced to develop as neural tissue. On the other hand, it seems that the presumptive mesoderm region does not necessarily retain its original polarity when it develops into mesoderm. No results have been recorded which suggest that a piece of graft mesoderm has developed with a polarity opposite to that of a larger mass of host mesoderm with which it is in contact, although of course the mesoderm in relatively isolated grafts may retain its original polarity since there is nothing to make it change. Certain cases of the reversal of mesodermal dorso-ventrality are clear. For instance, in Waddington & Schmidt's No. 32–136 (shown here in Fig. 40), the graft was originally made with reversed polarity and this must have been in turn reversed to give the degree of incorporation with the host which the specimen exhibits.

The dorso-ventrality of the endoderm has never been tested.

In the Amphibia, the mesoderm also shows a considerable readiness to become incorporated with that of the host, and in certain experiments (e.g. those of Mangold, 1925, and Bytinski-Salz, 1929) this has sometimes involved a reversal of dorso-ventrality. In the chick, we have seen that the development of mesoderm probably always involves the passage through a mesenchymatous phase, in which the cohesion of the cells is reduced, and this might be expected to provide a further reason why the polarity should be easily changeable. In Amphibia, Luther (1934) has shown that the dorso-ventrality of the ectoderm can be changed, at least up to the neurula stage, if small pieces are grafted inside-out on to the surface of other embryos. As Abercrombie points out, the chick experiments have not subjected the ectoderm of that form to any such definite influence, and the fact that no reversals have yet been effected should not be taken to prove that they are impossible; they might be possible if the right conditions could be found.

(b) *Medio-lateral organisation.* As in the Amphibia, an isolated neural plate induced in the chick has a strong tendency to be bilaterally symmetrical even when the inducing graft contained only one lateral half of the axis (Waddington, 1933a). This tendency is greatly modified when the induced neural plate lies near the host axis. The two plates then tend to become joined, and to form a single bilaterally symmetrical structure; examples are shown in Figs. 38, 40, and others are known in which the combined neural plates are comparatively symmetrical although the mesodermal parts may be highly asymmetrical. This situation is rather different from that found in the Amphibia, in which, when two neural plates lie side by side, each begins to fold up around its own median axis.

Rather few experiments have been made which would reveal the capacity for medio-lateral regulation of isolated pieces of streak which yield mesoderm, since most of such grafts have been bilaterally symmetrical. There is no doubt that the medio-lateral axis is still labile, since it must have been affected in all those cases in which a bilaterally symmetrical graft has been completely incorporated into the host's body. We have more information about the behaviour of asymmetrical regions of the streak in rather different circumstances, namely those in which a longitudinal split has occurred in a whole blastoderm, or in a large part of one, either in consequence of a longitudinal cut or owing to the absence of the median tissues in a posterior half (p. 80). Regulation in these circumstances is usually very incomplete (see Fig. 25 b), although Waterman (1936) has found signs of it in some cases in which a longitudinal cut was made anterior to the streak through the presumptive head region. But, as was mentioned earlier (p. 72), Abercrombie & Bellairs (1951) have shown that if the continuity of tissue is restored by grafting a piece of epiblast into the hole between the two sides of the streak, very complete regulation can occur, each side then forming a bilaterally symmetrical axis.

The pattern of the cross-section, even of a bilateral axis, seems to be partly dependent on tendencies inherent in the individual tissues and partly on mutual interactions between them. It has been found in many cases that when the notochord is absent, the somites tend to be united in the mid-line (Waddington, 1932; Wolff, 1936; Fig. 42), but it is likely that this is not a direct consequence of the absence of chorda, but rather that in these embryos the somitic mesoderm was also to some extent deficient. Similar effects have been described in amphibian embryos treated with lithium salts (e.g. by Lehmann, 1938), but in these, the effect might be attributed to injury to the presumptive somitic mesoderm rather than to the absence of a definite part of it. Something similar may have occurred in Wolff's cases, which were produced by X-irradiation of Hensen's node, since it is not certain whether the irradiated tissue is completely eliminated, or whether some of it remains alive in an injured condition. Waddington's specimens, however, resulted from the physical removal of parts of the node region, and there is no reason to suppose that any of the remaining tissue was unhealthy. The union of the somites would therefore seem to be a simple result of the absence of certain tissues. These tissues certainly include the notochord, and it might seem possible to argue that the somites fuse whenever the chorda is absent, as Holtfreter (1933) at one time suggested for the Amphibia. However, both Waddington (1935) and Wolff (1936) have drawn

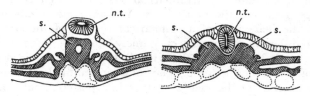

Fig. 42. Sections of embryos in which the somites (*s.*) have united in the mid-line, beneath the neural tube (*n.t.*), following X-irradiation of the node. (After Wolff.)

attention to cases in the chick in which the mesoderm has separated into two lateral sections even in the absence of the chorda. In Waddington's cases, which occurred in the posterior portions of blastoderms cut behind the node, the neural tissue was absent also, but development had not proceeded very far, and it is possible to argue that fusion in the mid-line would have occurred at a later stage. Unless this quite unsupported hypothesis is accepted, it is impossible to agree to the suggestion of Wolff, who attributes failure of the separation of the mesoderm into two lateral bands to an insufficiency of neural tissue. Wolff's evidence is easily explained by attributing the deficiencies in the neural tissue to those in the mesoderm, instead of vice versa; and the simplest explanation of the fusion of somites in the mid-line would seem to be that it is not primarily dependent on either the neural plate or the chorda, but occurs when, in the absence of the chorda which would otherwise act as a mechanical barrier, the median faces of the somites are missing.

Grünwald (1935) has described a number of naturally occurring malformations in which the neural groove (or parts of it) is abnormally wide, and has shown that this is always accompanied by an unusual width of the somitic mesoderm. The somites may be very irregularly shaped, and there may be several longitudinal series of them instead of only two, but Grünwald claims that they always extend exactly under the full width of the neural tissue. A similar result was produced when an agar plate was placed between the embryo and the vitelline membrane, in such a way as to prevent the folding of the neural tube (Grünwald, 1936b). He suggests that this extension of the somite zone is a secondary consequence of the abnormal width of the neural plate, which is supposed to become enlarged subsequent to its original induction by the mesoderm, and then itself to induce an unusually wide region of mesoderm to become segmented. Although such a reversal of the direction of inductive action cannot be considered in any way impossible, the evidence for it in this case is slender, and it seems almost equally plausible to suggest that the abnormality has arisen first in the mesoderm, which has then induced an expanded neural plate. It may also be that the mesoderm can become segmented into a quite regular series of somites in regions from which the neural tube is absent, for instance, in posterior half-blastoderms (p. 80, Fig. 26a).

It is well known that in the Amphibia, the shape of the cross-section of the neural tube is dependent on the nature of the surrounding mesoderm (cf. Holtfreter, 1933f; Lehmann, 1945), the tissue being thicker when it lies against somitic mesoderm and thinner over the chorda. The situation is probably similar in the chick, although there is still not very much evidence, owing to the difficulty of keeping embryos operated in vitro alive until a stage when such differences become well developed. In Wolff's specimens with fused somites, the floor of the neural tube was greatly thickened, and although this is what would be expected under the influence of the medianly placed mesoderm, it must not be overlooked that the effect might have been a consequence of a direct injury to the presumptive neural tissue by the irradiation. However, it has often been found that isolated pieces of neural tissue developing among mesenchyme in grafts give rise to completely circular tubes, as in the Amphibia, while neural tissue induced near a chorda often shows the appropriate thinning in its immediate neighbourhood. Cases are also known, however, of normally shaped neural tubes which extend through short regions from which the chorda is absent (cf. Waddington, 1938).

(c) *Antero-posterior organisation.* In discussing the antero-posterior organisation of structures we need to pay attention both to the direction of the axis (i.e. which is the cephalic and which the caudal end) and also the level of the axis represented (i.e. whether the anterior end represents the fore-, mid- or hind-brain, or some more posterior level of the neural tube). And we must consider not only what tendencies are originally present in a graft, but how far they are transmitted to the induction and to what extent they become modified by influences proceeding from the host.

By analogy with the Amphibia, one would expect that pieces of the streak representing different levels of the axis would have tendencies to develop into their presumptive fate and to induce corresponding regions. The evidence as regards self-differentiation of fragments of the streak has already been summarised (p. 82), and we have seen that the different levels differ in a manner to be expected, although there is some evidence of regulatory phenomena. There are also many cases in which an anterior fragment, grafted at some distance from the host axis, has induced the formation of a head; an example is shown in Fig. 39. The induction of posterior regions by posterior grafts cannot, perhaps, be demonstrated so spectacularly in the chick as in Amphibia, since in the former there is nothing so diagnostic as the tail fin. The evidence is thus largely negative in character, and consists in the fact that a recognisable head is never induced by the region posterior to the node, unless it lies close enough to the host's head to arouse the suspicion that host influences might have complicated the situation.

Isolated parts of the streak not only have tendencies to develop into certain levels of the axis, but the direction of the axis is also at least labilely fixed in them. This is usually most clearly expressed by the mutual relations of the neural and mesodermal tissues which develop; in isolated grafts, the former occurs in the initially anterior part of the graft. The axial direction, however, is not directly dependent on the relations between the two types of tissue, since a case has been described (Waddington, 1933 a, No. 32–43) in which a graft, that gave rise only to mesoderm and included no neural tissue, nevertheless induced a structure whose antero-posterior axis was opposed to that of the host embryo and must therefore have been transmitted by the graft (cf. Fig. 33). These initial tendencies of the organiser grafts are often greatly modified by influences proceeding from the host. In these mutual interactions, it appears that anterior regions tend to dominate over more posterior ones, regions near the axis have a stronger effect than regions farther away, older tissues are less easily altered than younger ones, and induced tissues more affected than grafted ones.

Thus we find that grafts of posterior portions, placed in the anterior region of the host, may succeed in inducing a head; an example is shown in Fig. 40, in which not only the induction but also the graft has developed into a much more anterior structure than would have been expected were it not for the influence of the host. The head in these cases is always at the most anterior end of the induction, as judged by the orientation of the host, and this is so even when the graft was made with the reversed orientation. Waddington & Schmidt (1933) have described a case (No. 32–217) in which the anterior half of the primitive streak was grafted near the host's head, with reversed orientation, and has induced a structure which contains a head with an orientation similar to that of the host although the orientation of the graft does not seem to have been changed. Older grafts, taken from head-fold or later stages and not extending to the anterior part of the head, do not induce heads even in the anterior region of the host (Waddington, 1933a). They are presumably less susceptible to modification by host influences.

Grafts of the posterior primitive streak, placed near the host's head, never show any signs of suppressing the anterior character of the host's development (except in so far as they may cause mechanical malformations) but anterior grafts in the posterior may convert the posterior end of the host's axis into a head (Waddington, 1934, No. 33–3). Grafts into the neighbourhood of the posterior end of the host's primitive streak, however, are frequently inhibited in differentiation, giving less neural tissue than would be expected, and they rarely induce very satisfactorily. In the specimen No. 33–3 just mentioned, the graft was originally placed just in the area opaca posterior to the end of the streak.

The greater effectiveness of the region near the axis is shown by the fact that when an induced neural plate is confluent with the host's plate, it is usually in the region of the junction that it exhibits the most anterior character, even if this involves a reversal in the initial orientation of the graft; an example is that illustrated in Fig. 32.

The mutual influence of host and graft was studied in fuller detail by Abercrombie & Waddington (1937), who made grafts of fragments of primitive streak under the streak of a host. They found considerable evidence of a tendency for the graft to become harmoniously incorporated into the host. This frequently involved the reversal of the direction of the axis in the graft, a change which takes place comparatively easily in grafts of posterior fragments into the anterior, but with rather more difficulty in presumptively anterior grafts. Thus, if an anterior graft is made with its original axis opposed to that of the host, it may suffer a reversal of its orientation, and then, if it comes to lie in the somite region of the host, it may develop head mesenchyme at its new anterior end in the neighbourhood of the host's somites. Its axis has then been reversed in direction but not changed in level. Changes in level, however, may also occur. Posterior grafts may, with or without accompanying reversal of orientation, be brought to develop some head mesenchyme, but such changes in level do not seem to succeed in eliciting the development of neural tissue from grafts which did not contain any presumptive neural material. In some cases, the changes in level are such as to bring the grafted tissue into almost exact conformity with the host; a good example is the specimen shown in Fig. 40, in which the fore-gut, heart and other organs, even including the inter-somitic grooves, are exactly parallel in host and induction. Often, however, the correspondence is much more rough and inaccurate.

Fusion of host and graft tissues sometimes occurs. It appears to be most easy in the head mesenchyme, which often forms a common mass even when the notochords and somites of the host and graft are completely distinct. Lateral fusion of neural plates formed in the same sheet of ectoderm is also common, for instance, between a host and an induced plate; but grafted neural tissue lying beneath a host plate very often fails to fuse with it. Perhaps indeed, it is misleading to speak of the fusion of host and induced plate, since it is more likely that what happens is simply that the induction acts over an area wide enough to cause the two plates to appear in contact. One frequently obtains the impression that, when an induced plate approaches rather near a host plate, the induction has, as it were, spread so as to cover the space which might have been expected to show between them; this might be due to the summation of the concentration of an evocator contributed both by the graft and by the host organiser.

The fusion of mesodermal tissues is difficult to establish with certainty, in the absence of any effective method of vital staining. Autonomous organs, such as somites or chorda, presumably derived mainly from the graft, are frequently found. They certainly show that somites do not necessarily tend to fuse unless kept apart by the presence of notochord or neural tissue, as suggested by Wolff (see p. 118). But it is not at all impossible that each of these separate somites, and possibly the separate notochords, are built from both graft and host cells. Moreover, cases can be seen in which there is a single somite, perhaps larger than normal, which gives every appearance of resulting from the fusion of the two elements; and it is certain that such fusion must have occurred in those cases in which the graft has been completely incorporated into the host's body, or is indicated only by an increase in the size of certain organs. Possibly in these cases, the fusion takes place before the differentiation into somites and notochord begins, for instance, at the stage of invagination from the streak.

The importance of early fusion for ensuring good incorporation and regulation has been made very clear by some experiments recently published by Abercrombie (1950). He excised various lengths of the primitive streak, and reimplanted them either in the original orientation (controls) or after reversal of the antero-posterior axis. Healing was often complete, and perfect embryos developed not only in the control series, but in those which had suffered reversal of parts of the streak. Regulation of this kind occurred in all the types of experiment; but, in those in which the node region of fully grown streak stages was involved, regulation was complete only if the reversed section was very short, not more than some 10 % of the whole streak length. Several types of incomplete regulation were found. In the simplest, a failure of healing led to formations similar to those found after mere removal of parts of the streak, such as the development of a split posterior axis. In other cases the median strip of the axis (neural floor and notochord) did not fail altogether but took the form of thin undifferentiated epithelium. In one of the most interesting types of incomplete regulation, there was a tendency for two complete axes to appear; similar duplications were found more frequently when posterior parts of the streak or non-streak epiblast were grafted in place of the anterior end of the streak (Abercrombie & Bellairs, 1951 and see p. 72).

These experiments are a powerful demonstration that, although, as we have seen, the antero-posterior organisation of the embryo seems to be strongest in the immediate neighbourhood of the axis, it cannot be wholly carried by the median strip which makes up the fully grown streak. Indeed, we see that the organisation of the regions outside the streak (presumably the main mass of the presumptive neural plate, some of the presumptive axial mesoderm and some already invaginated mesoderm) can overcome and reverse the orientation of the streak itself. The result is not entirely surprising, since essentially similar phenomena occur in the Amphibia (cf. Waddington & Yao, 1950), but they remind us that the organisation centre of the chick must not be too narrowly interpreted as being confined solely to the streak itself; it is certainly a larger region, which probably includes at least the whole presumptive axial mesoderm. It may be, moreover, that the reversal of orientation is effected by the whole region which possesses the dynamical tendencies to take part in the morphogenetic movements which occur as the streak regresses, and this region may well be larger than that which possesses inductive powers.

(d) *The nature of the individuating factors.* The processes and reactions by which a morphologically normal embryonic axis is formed, and which are involved in the mutual influences of host, graft and induction, were originally discussed by Waddington & Schmidt and Abercrombie & Waddington in terms of an 'individuation field'. This concept was introduced by the statement (Waddington & Schmidt, p. 555) that 'an organisation centre is surrounded by a field, which may be called the individuation field, within which its unifying action makes itself felt'. The use of the word 'field' has been frequent in experimental embryology since it was first popularised by such authors as Weiss (1930), Gurwitsch (1922) and others. It draws attention simultaneously to the spatial extent of a given process and to its unitary character. Both these characteristics are exemplified in a very clear way in the phenomena of individuation. But the field concept, as was well recognised even by its proponents such as Weiss (1939), carries with it the danger of tempting one to attribute causal efficacy to the field, by writing, for instance, that the individuation field suppresses the development of a certain tissue or induces another one; whereas the concept should be employed in a purely descriptive manner, the correct formulation being that in a certain part of the individuation field a given type of development is suppressed or encouraged. The causally effective factors must be sought for individually and found in agents of known effectiveness, not in the hypostatisation of a formal concept. The term 'individuation field', although it can be properly used if care is taken to employ it aright, has for these reasons not been introduced in the main body of this discussion.

We remain, then, confronted with the problem of solving the nature of the causal processes which bring about individuation. It would appear that, on the first level of analysis at least, they must be of many different kinds. It is clear that the assumption of a defined geometrical shape is in the first place a physical process which must be brought about by the operation of physical forces. These forces often seem to have a 'field' character, in that they operate in a graded way over a considerable area. For instance, there must be forces which bring about the folding of the flat neural plate into a neural groove and eventually a tube. It seems that these may spread in a more or less unbroken system over the whole area constituted by an induced plate which is fused with the host embryo; and thus we obtain the co-ordinated foldings seen, for example, in an embryo such as that shown in Fig. 40. Again, the forces which bring about the meristic segmentations of the axial mesoderm into a series of somites seem sometimes to operate as a system over a wider area, and to be able to cause a parallel segmentation of mesoderm not belonging wholly to the host. It is to forces such as these that we must attribute any complete incorporation of extraneous material into the host's body.

The nature of these forces is still almost entirely unknown in the chick, and little enough is known about them in the Amphibia. In the latter group, it is probable that changes in the relative properties of the cell-against-cell and cell-against-medium surfaces play an important role (cf. Waddington, 1942; Holtfreter, 1943, 1944), but this is perhaps less plausible in the birds, in which, owing to the small size of the cells, the cell-against-medium surface is relatively much less extensive. Some indication of the field laws of the somite-producing forces could perhaps be obtained from a study of cases in which the segmentation has taken place in a region which has been made abnormal by an induced curvature of the axis.[1] Some cases of curvature are available (unpublished material of Waddington), and one is figured here (Fig. 43). It is clear that the pattern of segmentation is altered, but no definite conclusions can be drawn from the small number of specimens; the subject seems likely to repay further study.

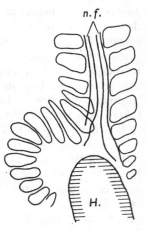

Fig. 43. Curvature of the row of somites round a hole (*H.*) in the mid-line of the blastoderm, following removal of the second quarter of the streak; *n.f.* neural folds.

It is important for the understanding of these individuation forces to realise that they are not only carried as inherent tendencies by isolated fragments of tissue, but can be transmitted by some form of induction process. For instance, Abercrombie (1937) has described how grafts of epiblast placed beneath the streak have had induced in them the capacity for backward elongation. The nature of this induction remains extremely obscure, but the fact of it cannot be denied. A similar remark, perhaps, can be made about the inductive reversal of the antero-posterior axis of a graft; but possibly we could explain this either as a consequence of the induction of a reversed morphogenetic movement, or of a sequence of regional inductions of the kind discussed in the next paragraph.

Individuation obviously involves something else besides physical effects on the anatomy of the tissues, since we have seen that changes in presumptive fate may occur, and tissues be produced,

[1] Attempts to bring about the formation of somites in a region in which two axes intersect at right angles have not yet been successful.

which would not have occurred in straightforward self-differentiation. There seem to be two main types of change. In the one, there is an alteration of level such that unexpected tissues are produced, which do not necessarily correspond to those in the neighbouring parts of the host. This may be called a 'regional induction', since the effect is to confer on the induced tissue the character proper to a certain region or level of the axis. In considerable contrast to this, there seems to be a tendency for a given type of tissue to induce the formation of similar tissue from available material in its neighbourhood, a phenomenon which was spoken of by Mangold & Spemann (1927) as an 'assimilative' or 'homoiogenetic' induction. The suppressive effect of the posterior end of the primitive streak may be an example of this; and the cases of good incorporation of a graft into the body of the host, or the formation of a single somite containing both host and graft tissue, would seem impossible to explain without postulating some such action, although a little weight can be allowed to the possibility of a fusion between host and graft previous to the time of determination.

It is no exaggeration to state that nothing is known with certainty as to the influences which produce regional or assimilative inductions in the chick. One of the most obvious questions to ask concerning the regional inductors is whether they differ only quantitatively, or qualitatively as well. Even in the Amphibia, no definite answer can yet be given (see discussion in Lehmann, 1945; Waddington & Yao, 1950), and still less is known in the chick. The process of assimilitative induction has also been little studied from the point of view of its physico-chemical mechanism. It is perhaps best known in the phenomenon of the induction of neural tissue by neural tissue (Mangold & Spemann, 1927). Such a process can give rise to a 'self-reproducing' tissue, provided a constant supply of young and competent tissue is made available. When similar phenomena are encountered in adult or nearly adult tissues, the intervention of a self-reproducing cytoplasmic particle (a plasmagene or cytogene) has been invoked (cf. Medawar, 1947). Such hypotheses are certainly not required in the embryonic case, even if they are in that of the adult; it is simpler to postulate that the competent cells contain an inactive complex from which an effective inducing agent can be easily released. This suggestion has been shown to be extremely probable in connection with the evocation of neural tissue in the Amphibia (Needham, 1942; Waddington, 1940; Holtfreter, 1948). It has not been so fully explored in connection with other tissues, but it appears to provide the simplest explanation of all assimilative inductions, particularly if we consider the process of breakdown to be of an autocatalytic type, which is encouraged by the presence of the free evocator.

CHAPTER VII

BIOCHEMICAL MECHANISMS OF
EARLY MORPHOGENESIS

THERE is, of course, an enormous body of data relating to the
biochemistry of the chick embryo. No attempt will be made in
this monograph to review the whole of it; summaries may be
found in the works of Needham (1931, 1942) and Brachet (1944).
We shall be concerned only with those investigations which have
revealed, or suggested, some close connection between the biochemical events and the primary morphogenetic phenomena.
The accounts of these constitute a much smaller corpus; in fact,
the chick blastoderm has not proved itself very suitable for studies
of this type, and our knowledge amounts to little more than that
yielded by comparatively few attempts to confirm conclusions
which were primarily based on amphibian material.

As far as the formation of endoderm is concerned, our first
information was provided by Jacobson (1938a). Woerdemann
(1933) had a few years previously produced evidence for a rapid
diminution of glycogen in the cells of the animal hemisphere of
the amphibian gastrula at the time when they became invaginated through the blastopore. His technique was called in question
by Pasteels (1936c), who claimed that only after the use of special
fixatives (preferably Bouin-dioxan) could one be certain that the
glycogen remained in its original location; the comparative lack
of glycogen in the invaginated cells found by Woerdemann was
attributed to a fixation artifact. However, Heatley & Lindahl
(1937), using a microchemical method, showed that there was
a rapid decrease of glycogen in these cells, though this did not
lead to the complete loss which Woerdeman described. Jacobson
was led to apply the same line of thought to the chick embryo,
and to inquire whether gastrulation was associated with a rapid
consumption of glycogen in that form also. He employed several
histochemical methods, including those recommended by Pasteels,
and claimed that glycogen was rapidly lost from the endoderm
cells which he described as being invaginated from an endoderm-
blastopore situated in the posterior region of the blastoderm.
This mode of origin of the endoderm has not been accepted by

subsequent authors (p. 12), and it remains obscure how much significance can be attached to the alleged glycogenolysis. As Brachet (1944) points out, normal biochemical tests have failed to reveal the presence of any glycogen in the blastoderm at this stage.

Brachet (1940) has employed cytochemical methods to trace the location of certain proteins, in particular those which stain with basic dyes such as pyronin and those which give the nitro prusside reaction (following the technique of Giroud and Bulliard) after denaturation with trichloracetic acid. The latter become visible because the denaturation gives rise to free —SH groups, and Brachet gives reasons for supposing that both techniques in fact reveal ribonucleoproteins. In the unincubated chick egg, both methods disclose a greater concentration of such substances in the upper layer, while the hypoblast, by the time it is definitively formed, is almost free of them.

Considerably more work has been done on the primitive streak stage. Jacobson (1938b) carried his histochemical studies of glycogen distribution into this phase, and claims that there is a further period of glycogenolysis associated with the invagination of the mesoderm through the primitive streak. He also made a study of the distribution of lipoids. He found that compounds staining in lipoid stains (such as sudan orange, scarlet R, etc.) increase in concentration in the epiblast cells as these move towards the streak, but decrease again after invagination, when the cells are moving towards the sides in the form of mesoderm. Only in the forwardly growing head process, the lipoid content remains high for a short time, until differentiation into notochord starts, when there, too, it decreases. Jacobson claims that these changes in lipoid content materially affect the movements of marks made with vital stains such as nile blue, and bases on this much of his criticism of the ideas of Pasteels concerning the morphogenetic movements of gastrulation (p. 24). He also made some further tests to identify more precisely the type of lipoid involved, and came to the conclusion on the basis of the Smith-Dietrich and Ciaccio methods, that they were probably sterols or related substances.[1] It was already known from the work of Waddington & Needham that such compounds are very

[1] For a review of the reliability of histochemical tests for lipoids, see Cain (1950).

active neural evocators in Amphibia, and it had been suggested that the naturally occurring evocator might be related to them; Jacobson, therefore, suggested that the changes in lipoid content which he had demonstrated might be the visible signs of a process intimately connected with the activity of the streak as an inducing agent. The suggestion is an interesting one, and should be followed up. Rather surprisingly, we seem to have no information on the distribution of similar substances in the amphibian gastrula.

Jacobson also found that, at the time when mesoderm invagination ceases in the anterior part of the blastoderm, a region lying around and anterior to the node stains more deeply in fat stains and also shows an unusually high concentration of necrotic cells. This region he supposes to be the presumptive neural plate, a conclusion which is supported by Spratt (1947 b) who described a similar condition. At this stage, induction of the neural plate by the underlying mesoderm has presumably begun, but the presumptive neural tissue is certainly still labile; the concentration of lipoids in it, if it is a fact, must be one of the very earliest stages in its differentiation into the neural plate.

Brachet (1940, 1944) has also applied his methods for sulphydryl and ribonucleoproteins to the primitive streak stage. The greatest concentration of these substances is found in the primitive streak itself, and in the mesoderm formed from it, but they also make their appearance quite soon in the neural tube. In the early somite stages, there is a fairly well-marked cephalo-caudal gradient, the anterior region having the greater concentration. Brachet is inclined to attribute great importance to such basophilic substances in the process of induction. Gallera & Oprecht (1948) repeated Brachet's observations in somewhat greater detail, and found that in the non-incubated blastoderm no basophilic substances of the ribonucleoprotein type could be detected. They did not appear in the endoderm till quite late, but seemed to increase fairly rapidly in concentration in the epiblast as the streak formed. These authors confirm the existence of a cephalo-caudal gradient in the mesoblast, and state that a medio-lateral gradient is also apparent as invagination proceeds, but this disappears in neural plate stages, when the whole mesoderm is fairly evenly coloured by basic dyes, with perhaps some concentration around the developing blood islands and in the chorda. By this time, basophily is very marked in the neural plate itself, but is lost from the extra-neural ectoderm, except at the outer border of the area opaca, where rapid growth is going on. Gallera & Oprecht describe particularly strong concentrations of basophilic substances at the contiguous faces of two tissues which come together in an inductive relationship as, for instance, the optic cup and lens ectoderm, and they support Brachet's ideas as to the importance of ribonucleoproteins in induction.

Stockenberg (1937) has also devoted some attention to the location of concentrations of somewhat similar materials, which were identified by their stainability with vital dyes, particularly neutral red. He suggested that they are regions either of specially active yolk-absorption or of extensive cell degeneration.[1] He found them to be most strongly marked in rather later stages (2–3 day embryos) in which they seem to be characteristic of places at which folds of tissue are becoming fused with one another, as for instance in the closure of the neural folds, the sealing off of the lens or ear vesicles from the epidermis, the junction of the amniotic folds, etc.

An antero-posterior gradient, similar to that described by Brachet, had already been noticed, in a preliminary way, by Rulon (1935). This author was primarily concerned with the demonstration of a gradient of oxidation-reduction potential, to which Child attributed such importance in his well-known theory of 'axial gradients'.[2] Rulon stained the embryos with oxidised janus green, placed them in conditions of low oxygen, and watched the change of colour from blue-green to bright red which is brought about by the reduction of the oxidised dye. He found that, in the primitive streak stage, the change proceeds fastest in the streak itself, there being also a cranio-caudal gradient from the node down to the posterior. As the brain develops, a second centre of activity is formed in the anterior of this, so that there is a double gradient, from the tip of the brain posteriorly, and then from the node posteriorly. Similar gradients were found by Hinrichs (1927), Buchanan (1926) and Hyman (1927) who subjected the blastoderm to deleterious agents, the former to ultra-violet irradiation and the two latter to cyanide. The inter-

[1] For a general review of the incidence of cellular degeneration during development, see Glücksmann (1951).
[2] Cf. Child (1946) for a review of this theory in relation to organiser phenomena.

pretation of such experiments has been extensively discussed by Needham (1931), who points to a number of reasons why they cannot be accepted as demonstrating the existence of a gradient in respiratory rate, as their authors have claimed. Direct measurements by Philips (1941, 1942) using the Cartesian diver micromanometer have failed to demonstrate any differences in oxygen consumption in the different levels of the streak and head process; but the pieces compared included large areas of area pellucida on each side of the streak, so that even quite large differences in respiration of anterior and posterior regions of the streak might have escaped detection.

Probably more importance should be attached to studies in which definite enzymes have been located. Moog (1943 a) showed that an indophenol oxidase (probably cytochrome oxidase) is present on the first day of incubation, and that in the streak stage it is also distributed in a gradient, i.e. is most concentrated in the anterior part of the streak. It remains entirely obscure, however, whether this is merely another reflection of the well-known anteroposterior law of development, or whether the presence of the enzyme has any causal significance. Similarly, there are no data relating to the chick which would enable one to interpret the gradient in pyronin-elective proteins described by Brachet, and it is primarily on the conditions in the Amphibia that he bases the suggestion, discussed in his 1944 book, that the substances concerned are intimately connected with induction.

The location of another set of enzymes, the acid and alkaline phosphatases, has been traced in some detail by Moog (1943 b, 1944). They are both present in most tissues of the early blastoderm, but their distribution cannot be explained on the hypothesis, which is supported by some authors, that they are characteristic of all rapidly proliferating tissues, since they are absent from some of these, such as the extra-embryonic blastoderm during its expansion over the yolk. Moog suggests that in interpreting their location, it is necessary to distinguish between two different categories: on the one hand, concentrations of enzyme in particular differentiated tissues, where they can be regarded as part of the specific histological character of the cells, and, on the other, concentrations occurring in a primitive phase, before full differentiation has taken place. It is the second category which is of main interest here. Moog finds that phosphatase (mainly alkaline) tends during this primitive phase to be most in evidence between the time of determination and the completion of differentiation. For instance, it does not become strongly evident in the hinder part of the body till the second day, and again it appears in the liver diverticulae between their first appearance and their definitive differentiation. Moog suggests that the enzyme may be concerned, during this primitive phase, with the synthesis of specific proteins, on which the visible histological differentiation essentially depends.

A different, and perhaps ultimately very promising, method of studying the formation of specific proteins during development is by the use of immunological methods. Waddington, in a few preliminary and unpublished experiments in 1938, cultivated chick embryos in a medium made up with plasma from adult rabbits which had been given a course of injections of 7-day chick embryo brain. No effective immunity was manifested against the 1- or 2-day-old embryos, which developed quite normally. More elaborate studies by Schechtman (1947) show that antisera prepared in rabbits against adult fowl serum will react with saline extracts of blastoderms from the streak to the early (15–17) somite stages, but that the reactivity is lost after thorough adsorption of the antisera with yolk. He concludes that the non-vitelloid constituents of adult serum are not present in the young embryo of these stages, and gives some evidence that they begin to make their appearance from the sixth day onwards, one at least being definitely detectable by the fifteenth day.

Immunological methods have also been used to detect the presence of organ-specific antigens in an attempt to determine the time of their first appearance. Burke, Sullivan & Weed (1944) claimed that the antigens present in adult fowl lens cannot be detected in the embryo of less than about 6 days (144 hours), and therefore argued that the organ specific substances appear only after considerable morphological differentiation. Using more delicate methods, however, ten Cate (1949) and ten Cate & Doorenmaalen (1950) have been able to demonstrate a reaction between adult lens antiserum (prepared in rabbits) and embryonic lenses at the stage when the lens vesicle is not yet fully free from the epidermis (about 60 hours). At this stage the morphological changes are confined to the folding, and the special characteristics, differentiating the lens from other ectodermal placodes, have not yet appeared. It therefore seems likely that the chemical differentiation of the lens occurs before the characteristic morphology is assumed. It would indeed be surprising if this were not so, since the morphological changes must surely be the consequences of some underlying change in the nature of the substance composing the cells; such changes of course become visible to the histologist shortly after the 60-hour stage, by the appearance of lens fibres.

Ebert (1949, 1950) has also shown, by absorption and precipitin tests, that antisera prepared in rabbits against adult chick brain, heart or spleen will react with extracts from primitive streak or early somite blastoderms, which must therefore contain antigenic substances closely related (or identical with) those of the adult organs. When blastoderms were cultured in media containing the antisera, non-specific lethal and growth-inhibitory effects were found at some dilutions and, oddly enough, these effects could not be obviated by absorption with organ extracts. At critical concentrations, however, some specificity in effect was noticed. In particular the anti-heart and anti-spleen sera had a more strongly inhibitory effect on the development of mesoderm than on that of brain, while this situation was not so marked or even reversed in the cultures in anti-brain serum. There were probably also differences between the anti-heart and anti-spleen sera, the former suppressing the heart differentially. Similar, though less complete results, were reported by Pomerat (1949) who treated blastoderms *in ovo* with anti-spleen serum.

Results of a very interesting but rather different kind have been reported in a preliminary way by Weiss (1947). Antisera were prepared by injecting autolysed adult chicken liver, kidney, or pectoral muscle into guinea-pigs, and these sera then injected into eggs of various ages between the third and eighth day of incubation. Weights were taken of the organs of the twentieth-day embryos. All antisera tended to produce some depression in weight, but this was less marked, or even reversed, in respect of the organ for which the serum was specific; thus the livers were larger in embryos injected with liver antisera than in those which had received the kidney antisera, and the kidneys larger in the latter group than in the former. Weiss, therefore, suggests that, superimposed on a general inhibition of growth brought about by any antiserum, there is a specific stimulatory effect on the homologous organ. This experiment, if it can be confirmed, is one of the first successful attempts to influence directly the fundamental developmental process of specific protein synthesis. Many more such attempts are likely to be made in the future, and their full understanding must await further work; the hypothesis of Weiss, for which his paper must be consulted, is by no means the only one possible.

Comparatively few attempts have been made to discover the morphogenetic importance of enzyme systems in the chick embryo by observing the effects of so-called specific inhibitors. Waddington (quoted by Needham, 1933) cultivated embryos *in vitro* on media to which KCN had been added and found that development proceeded relatively normally up to a concentration of at least 0·001M, which should reduce the normal respiratory intensity by at least 80 %; it is, however, by no means certain that this concentration was maintained throughout the life of the culture. In the same unpublished experiments, it was found that concentrations of sodium fluoride up to 0·005M had little effect. These results are similar to those of Brachet (cf. 1944, p. 392) on Amphibia. Relatively normal development also occurred in concentrations of lithium carbonate up to 0·01M, and nothing comparable to the specific effects of lithium salts on the amphibian embryo was detected.

The effects of asphyxia, produced for instance by waxing the shell, have been extensively studied from early times, but the results are hardly specific enough to suggest any conclusions as to the biochemical mechanisms of determination and morphogenesis. Demuth (1939) has carried out the converse experiment, incubating eggs in an atmosphere considerably enriched with oxygen. He observed an inhibition of general growth, but the degree of differentiation of certain organs, such as the eye, epiphysis, hypophysis and thyroid, was in advance of that of normal controls and very considerably in advance of that characteristic of the weight which the embryo had attained. Demuth attributes the effects to a shifting of the metabolism away from glycolysis towards an oxidative respiration, but this requires further study.

Very many experiments have, of course, been made dealing with the application of comparatively non-specific deleterious substances to the embryo, but in most of these the substances have been merely injected into the egg and their effects investigated at a late stage of development. The extensive literature has indeed little to contribute to analytical experimental embryology; an introduction to it, however, can be gained through the articles of Landauer (1941c) and Grünwald (1947a), who also deal with the effects of deficient maternal diet (see Cruikshank, 1931). Some of the most interesting of such 'chemical injuries' are those produced by bacterial toxins (e.g. meningococcus, Buddingh & Polk, 1939; tetanus toxin, Gray & Worthing, 1941) and viruses (influenza A and mumps, Hamburger & Kabel, 1947), both of which have been shown to produce profound morphological effects during the first few days of incubation.

Of all general and unspecific agents, temperature is perhaps the easiest to control, and very many authors have applied shocks, either of high or low temperature, or have incubated eggs at unusual temperatures. In general, little of morphogenetic interest has emerged. A common reaction of the early stages is the formation of an 'anidian' blastoderm, i.e. one in which all trace of embryo or area pellucida is lost, the blastoderm consisting of a growing plate of area opaca, within which blood islands may develop; in fact, in some cases in which the disappearance of the embryo has taken place at a fairly late stage, the extra-embryonic blood vessels continue expanding until the blastoderm covers the entire yolk. In others a hole appears in the position of the area pellucida and development ceases at an early stage. The fullest account of such embryos appears to be that of Grodziński (1934b). Brachet (1949) claims that by high temperature shock it is possible in Amphibia to suppress the mobilisation of nucleoproteins in the blastopore lip region, and thus to inhibit the development of the embryonic axis. Attempts by Deuchar (not yet published) to repeat this on the chick did not yield such a clear-cut picture, but showed that the primitive streak period is more sensitive to incubation at 45° C. than the earlier stages. The effects seemed to bear mostly on the proper carrying out of the morphogenetic movements involved in gastrulation, and the abnormalities of the resulting embryos seemed to be related rather to the distortion or inhibition of these movements than to result from a more direct influence on the process of induction.

Tazelaar (1928) applied a temperature gradient either across the blastoderm from side to side or from anterior to posterior. The main effect was on the area vasculosa, which became considerably larger on the heated side. There was also a slight enlargement of the organs of the embryo itself on this side, and the somites on the heated side were shifted anteriorly by the distance of about the length of a somite.

Burr and his associates (Northrop & Burr, 1937; Burr & Hovland, 1937) have attributed an important role in morphogenesis to electric potentials which, they argue, are bound to be produced by metabolising tissues. By measurements in the neighbourhood of the chick embryo they claim to have demonstrated the existence of gradients in potential, the anterior being always positive towards more posterior regions, and the potential difference increasing steadily during at least the early part of development. The reality of the potential differences cannot, perhaps, be regarded yet as securely demonstrated, and their effectiveness is still more open to question. For instance, Gray (1939) has reported an extensive series of experiments in which an attempt was made to modify morphogenesis by the application of a direct current between electrodes placed on each side of the blastoderm *in ovo*. Abnormalities were duly found. But in all cases these could be exactly paralleled by the application of heavy metal salts, and their frequency was not related to the intensity or duration of the applied currents, which often allowed of normal development even when of just sub-lethal intensity. Gray concludes that none of the observed effects was due to the currents and that it is, therefore, extremely unlikely that any potential differences which may exist between different parts of the embryo can have any morphogenetic influence.

The curious reader will find a discussion of the role of mitogenetic rays in chick embryo development in Sorin (1928). Inhibition of development, which may lead to the death of the embryo, may of course be produced by a sufficiently heavy dosage of X-rays (e.g. Strangeways & Fell, 1927) but the results are not of much interest from the morphogenetic point of view.

Some consideration has been given by Hobson (1941) to the importance of other physical factors for chick morphogenesis. He examined the birefringence of early embryos about the time of neural tube formation and showed that the tube itself is birefringent, the sign being positive with respect to the long axis of the radially arranged cells in its walls. The birefringence can probably be attributed mainly to form birefringence of the plasma and nuclear membranes of the cell; but some persists even after dehydration and fat extraction, so that there may be some intrinsic birefringence of the cytoplasmic constituents, and the structure causing this may take part in the morphological changes associated with tube formation. Hobson also points out that dehydration causes flattening in a neural groove which has not yet closed into a tube by the fusion of its dorsal edges, and adduces this as support for Glaser's (1914) theory that the folding of the groove is due to differential imbibition of water by the ventral surface.

Davis (1944) has investigated the same problem by means of ultra-violet irradiation. He irradiated blastoderms in the stages from the primitive streak to the 8-somite embryo with ultra-violet rays of 2483–3130Å., and then incubated the eggs for 30 hours. Certain doses were found to inhibit neural tube formation, while permitting cell division and volume changes to continue normally; a small dose, about 0·01 ergs/mm.^{2}, inhibited closure over 50 % of the length of the nerve tube. By using monochromatic irradiation, photochemical efficiency curves were arrived at, which Davis claims to be extremely similar to absorption curves typical of sterols. He suggests that the latter are therefore likely to be involved in the morphogenesis; and it has already been noted (pp. 115, 131) that similar substances are extremely effective evocators in Amphibia and have some such action in the chick.

CHAPTER VIII

THE DEVELOPMENT OF SOME ORGAN
SYSTEMS IN LATER STAGES

The early development of the head

A considerable number of experiments have been made on the development of the head, and of the separate organs which it contains—the fore-, mid- and hind-brain, the optic cups, the lens, the nasal and auditory placodes. It is only in later stages that these organs are sufficiently separate to be considered individually, and the first part of our discussion must to some extent, deal with the head as a unit.

It has been pointed out earlier (p. 71) that in the conditions of *in vitro* cultivation, which, relatively speaking, are mechanically stable, considerable regeneration may occur after defects made in the primitive streak stage. A series of experiments in which defects were made just anterior to the node (i.e. to the presumptive head region) were described by Waddington & Cohen (1936) and further work of a very similar nature was published by Spratt (1940). In the former series, the defects were made by the removal of small triangles of tissue; the apex of the triangle, the angle of which varied between 30° and 90°, was at the level of the node, and the remainder of the triangle extended nearly to the edge of the area pellucida. When operations were made in the primitive streak stage, repair was often found to be complete before the appearance of a definite neural plate, and in these circumstances a fully normal head was developed (Figs. 44a, b). This occurred both when the triangle was medianly placed, or when it was somewhat to one side, so as to remove the presumptive tissue of the optic cup. Exactly similar results were obtained by Spratt, who removed rectangular fragments, which were wider immediately anterior to the node than Waddington & Cohen's, but which did not extend so far anteriorly.

The repair in the pre-neural plate stage occurs by the growth of the cut eges from all sides in towards the centre of the hole. In the embryos in which the defect had a narrow triangular shape, the healing was, therefore, mainly by a transverse movement of the two sides towards the mid-line. A failure of complete healing in such cases produces an embryo which is partially split along the mid-line. If, on the other hand, the wound extends transversely across the region just in front of the node, it tends to be repaired mainly by forwardly directed growth; and any insufficiency of this produces embryos in which the fore-brain is shortened. The results of these two types of failure are thus rather different, and must be mentioned separately (Figs. 44c, 45).

The medianly split embryos occurred only in Waddington & Cohen's series and were not seen by Spratt. As might be expected

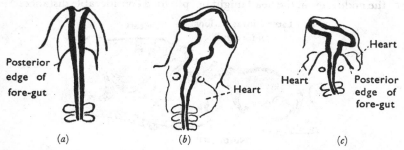

Fig. 44. *a* A triangle, with its apex (of about 60°) at the node, and its base near the edge of the area pellucida, was removed from a duck embryo with a medium primitive streak. After twenty hours' cultivation *in vitro*, the wound is repaired, and the head more or less completely regenerated. *b* A chick embryo from which a narrower triangle was removed from anterior to the node of a long primitive streak; regeneration of the head is complete. *c* A 60° triangle was removed from anterior to the node of a long primitive streak embryo, and the edges of the wound stretched apart; repair has been by forward growth from the apex of the wound, and the head is shortened. (After Waddington & Cohen.)

from what we know of the map of presumptive areas, the floor of the neural system is affected over a longer stretch than is the roof. The developing neural tissue in the defective region does not, as might have been expected, heal together with the neural material of the opposite side, to give closed, though defective, tubes. On the contrary, wherever there is an absence of the appropriate presumptive tissue, the tubes apparently remain unclosed, either dorsally or ventrally. The unhealed ventral edges usually come to lie in the neighbourhood either of the ectoderm of the head-fold, or of the endoderm of the developing fore-gut, and in spite of the difficulty with which they seem to coalesce with other neural tissue at this stage, they frequently become united with one or other of these comparatively foreign tissues (Fig. 45). The head-fold ectoderm and the fore-gut endoderm also frequently heal together; and union between neural plate and endoderm is common in more posterior regions in which the floor of the neural groove is lacking: for instance, in posterior segments of the blasto-derm (p. 80). We shall see that neural tissue of the stages later than the appearance of a well-developed groove is well able to heal together with its like (p. 146).

In cases in which the apex of the defect reached deeply into the node region, the head might be split for a considerable distance.

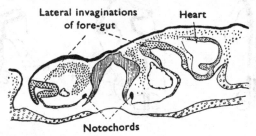

Fig. 45. Section through the neck region of an embryo from which a 45° triangle was removed anterior to the node. Partial regeneration, with defective floor of the neural tube, laterally displaced fore-gut and double heart (only one seen in section). Neural system is shown by vertical lines, endoderm, by small circles. (After Waddington & Cohen.)

It is then pertinent to ask whether both sides show any tendency to regulate to form a complete head. Waddington & Cohen found that the neural tube might show a tendency to become rounded up so as to form a tube with a closed cross-section; but they saw no sign of the appearance of a regenerated optic evagination on the defective side, which would indicate a real regulation. Waterman (1936), in a short communication, claims that this may occur in blastoderms in which the region anterior to the node has been split by a longitudinal cut no tissue being actually removed; and we have seen that in more favourable conditions, when other tissue is provided to heal the cut edge, Abercrombie & Bellairs have found that a longitudinal half embryo may regulate completely (p. 72).

In embryos in which the defect to the presumptive head region is disposed transversely, complete regeneration is also possible, if the wound is completely healed before the appearance of the neural plate. Even when the healing is not so perfect, Waddington & Cohen found that reorganisation might take place within the available material, so that a head is formed which possesses all the essential parts of the fore-brain, although shortened in length (Fig. 44*c*). This did not seem to occur in Spratt's material, and, according to his description, it certainly does not occur when, instead of a small defect being made, the whole anterior part of the blastoderm is removed by a cut slightly in front of the node; the defect is then not made good, and a part of the fore-brain is missing. A similar failure of regeneration in such circumstances was noted by Waddington (1932).

When defects are made in slightly older stages (head process or head-fold stages), both sets of authors find that regeneration proceeds very much less successfully. Spratt made a careful study of the situation when the anterior part of such blastoderms is separated from the posterior by a transverse cut at a definite level, and he showed that the two fragments develop in an essentially mosaic manner into the two complementary regions of the brain.

Defects of a less drastic character than the total removal of tissue have been made by Wolff (1934*c*, *d*, 1936). He found that if the major part of the blastoderm anterior to the tip of the head process is irradiated with a comparatively light dose of X-rays, a certain number of the embryos develop with a single median eye. Such cyclopia is, of course, well known in many classes of vertebrates, and seems usually to be due to a weakening of the underlying mesoderm rather than a direct effect on the presumptive eye tissue itself (cf. review of Mangold, 1931). The exact mode of operation of Wolff's irradiations remains uncertain.

The origin of the optic cup

In connection with his operative defects in head-process and later stages, Spratt observed good development of optic vesicles from small isolates taken from in front of the head process, which included only those parts which were removed in the regeneration experiments described above. Slightly less frequent and perfect differentiation was obtained from isolates taken from in front of the node in fully grown streak stages. In isolates from these stages, of course, a certain amount of mesoderm is included as well as the optic ectoderm. The situation is more or less comparable to that of an isolate from a late amphibian gastrula in which archenteron roof had been taken with the overlying ectoderm. In the chick it has been impossible to make experiments similar to those of Mangold (1928a, 1929a), in which capacity of the presumptive optic ectoderm was tested after isolation from the mesoderm. In the amphibian experiments the earliest isolates which gave rise to eyes were from middle yolk-plug gastrulae, a stage which would, perhaps, correspond to the anterior region of a not quite fully grown primitive streak. In amphibian isolates containing mesoderm, eyes can be obtained from much earlier stages. For instance, Holtfreter (1938) has found them in fragments taken from beginning gastrulae; and it is to be noticed that from these stages they were developed, not in the isolates containing solely the presumptive optic ectoderm, but, on the contrary, from those consisting of presumptive mesoderm; moreover, this mesoderm was not necessarily that whose normal fate is to underlie and induce the optic rudiment, but might be quite foreign mesoderm in which eyes were developed by a paragenesis both in respect of the production of ectoderm and in respect of the regionality of the structures formed.

It is very necessary to bear these facts in mind when attempting to evaluate the studies which have been made on the development of eyes in isolated fragments of the chick blastoderm. Thus Hoadley (1926a) obtained eyes in isolates on to the chorio-allantois from the stage of the short broad streak; and Murray & Selby (1930a) described a single case from a graft of an entire unincubated blastoderm. These results tell us nothing as to the epigenetic situation within the presumptive eye rudiment. Hoadley laid stress on the fact that more complete and more frequent differentiation occurred in isolates from later stages than in those from younger ones, and on this based a theory of gradually increasing 'segregations'. However, it is a very general observation that the longer the course of development which an isolate is called upon to perform in rather abnormal conditions, the less perfect the end-result produced, the earlier stages being particularly sensitive, and there seems no reason why such considerations should not be the explanation of Hoadley's results.

More recent workers with the chorio-allantoic method have attempted a more precise localisation of the region from which eyes may be obtained. Hunt (1931a) and Stein (1933) placed this area at, and just anterior to, the node at late streak stages, and between the node and a point somewhat anterior to the tip of the chorda in head-process stages. Clarke (1936) defined the area more closely, showing it to be concentrated near the tip of the head process, the node level at that stage no longer yielding eyes (Fig. 46). It must be pointed out that the two eye-forming regions demonstrated by this author do not comprise the same

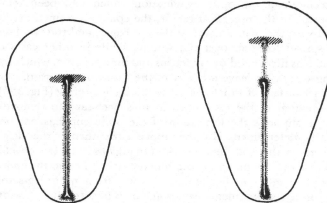

Fig. 46. Regions from which optic tissue is obtained in chorio-allantoic grafts (indicated by hatching) in streak and head process stages. (After Clarke.)

actual material, since in the streak stage the eye region of the head-process blastoderm would lie some distance anterior to the node. This strongly suggests that the isolations, at least from the streak stage, are not in fact testing the state of determination of the presumptive eye region, but that the eyes formed from the node level are the result of a paragenesis comparable to that by which fragments of early gastrula mesoderm gave eyes in Holt-freter's experiments.

The later development of the eye is considered on p. 151.

Later stages of the brain

Waddington & Cohen (1936) and Spratt (1940) both carried their studies on the regenerative capacities of the head into the early somite stages. The defects made by the former consisted in the removal of one lateral half of the fore-brain.[1] They found that

---

[1] Experiments in which only part of the presumptive optic evagination was removed will be considered later.

from the 5-somite stage till at least the 25-somite stage, a half fore-brain can become remodelled to form a closed tube. No optic evagination appeared on the injured side, but this side could induce the formation of a nasal placode from the epidermis which grew over the wound (p. 158). They, therefore, concluded that the paragenesis involved the regeneration of a complete fore-brain from the part which had been left intact, but not the replacement of the optic evagination which had been totally removed. In the process of repair, the epidermis sometimes first united with the neural tissue at the cut edges, and the two layers then spread together over the wound, while in other cases the wound was first closed by epidermis and mesenchyme, which were arranged so as to leave a cavity of the shape of the brain, along the inner surface of which the neural epithelium grew (Fig. 47 b).

The extent of the region which it is necessary to remove in order to suppress the formation of the optic cup has not been precisely determined. In the above experiments, the defects extended to the mid-line. Clarke (1936) observed the development of eyes from pieces of 'brain floor' isolated on to the chorio-allantois from 4- to 8- somite, but not from 9- to 12-somite embryos. These fragments were some 0·06 to 0·10 mm. in width, and thus presumably contained enough of the optic evagination to allow regulation to take place as it does when the distal part of the optic lobe is removed (p. 152).

Most of Spratt's experiments consisted in the complete removal of the fore-brain. After this operation, there was no replacement of it from the mid-brain, which simply healed up anteriorly. Some regulation occurred, however, when only the anterior part of the fore-brain was removed. Spratt also mentions very shortly (1940, p. 192) some experiments in which lateral halves of the fore-brain were removed, and which he states gave results in accord with those of Waddington & Cohen; however, the absence of the optic evagination apparently makes him unwilling to consider these as representing a regeneration of the missing parts, but he seems to have overlooked the significance of the fact that the missing side was sufficiently restored for its nasal placode to appear.

It has been observed by many students of regeneration during embryonic development, that the body can be analysed into regions such that if a complete region is removed it is not replaced but that if a part of it is left, that part can regenerate the whole (cf. Weiss, 1939). It would appear that we are confronted with a similar situation in the chick brain, the fore-brain being one such region, and the optic evagination another separate one. We shall see (p. 152) that although the optic region as a whole cannot be replaced, defects within that region are easily restored. We find, then, a rather high capacity for regeneration in these anterior head structures in early somite stages, limited only by this division of the whole complex into two more or less independent regions. This is perhaps in unexpected contrast to the situation in the earlier head-process and head-fold stages, in which regeneration capacity seems to be at a minimum. Such phenomena however, are, by no means unknown in other fields. Svetlov (1934) has emphasised, on the basis of experiments on the anuran tail bud, that the critical period of determination and of the beginnings of differentiation is frequently a time of minimal capacity for paragenetic teleogenesis. In the Amphibia, as in the chick, the capacity for regeneration in the nervous system seems to be at its least in the stage of the open neural plate (cf. Spemann, 1938, and Detwiler, 1936).

The results of some defects made to more posterior regions of the brain have been recorded by Waddington (1936a). The operations, which were made in connection with a study of the auditory placode, consisted in the removal of the dorsal half of one side of the wall of the neural tube between the level of the first somite and the mid-brain region, in embryos with between one and nine pairs of somites. It was again found that in favourable cases very considerable repair occurred, so that the dorsal ganglia (particularly the acoustico-facialis) might be normally formed. In other cases, however, in which repair of the neural tube did not proceed so successfully, this ganglion might nevertheless appear on the injured side, but was derived from neural crest cells of the opposite side, which had migrated across the dorsal surface of the tube.

Bilateral extirpations of the neural crest, both in the cervical and trunk regions, have been made by a number of authors (Jones, 1937, 1939, 1941, 1942; Yntema & Hammond, 1945, 1947; Hammond & Yntema, 1947; Hammond, 1949; Levi-Montalcini, 1947a; Levi-Montalcini & Amprino, 1947) and have been followed by absence of the corresponding spinal ganglia, although Levi-Montalcini (1947b) found that the sympathetic ganglia are formed (presumably from mesodermal mesenchyme) even after total removal of the neural crest in the neighbourhood. These regions of the crest also contribute to the cartilages of the head, and defects of the hyoid were found by Yntema (1944) to follow extirpation of the crest of the posterior cranial region.

Defect experiments of quite a different nature have been made on the heads of embryos belonging to these stages by Wolff (1934c, d, 1936). This author employed localised irradiation with X-rays to kill defined regions of the embryo. The exact fate of the injured tissue is not quite clear, since the operated embryos were kept alive, in the shell, as long as possible, and usually were not investigated till several days after the operation. It appears, however, that this tissue did not simply disappear and leave an open wound; but it does not seem to have played any part in later development, and further it does not seem to have been replaced to any large extent by a paragenesis of the remaining healthy tissue. The operations, therefore, cause the absence of regions which seem to correspond closely with those whose rudiments were injured. In this way, Wolff was able to record success in the goal which he had set himself, namely, to produce by artificial means certain recognised types of teratological mal-formation, such as cyclopia, otocephaly, etc., which lack the region between the eyes, the derivatives of the upper or lower jaw, etc. From the point of view of developmental mechanics the experiments do little more than demonstrate that under certain circumstances the different regions of the somite stage embryo may develop in any essentially mosaic manner.

Filogamo (1950) removed one eye-cup from embryos of 48–52 hours and investigated the effect on the differentiation of the contra-lateral optic lobe. This was found to develop quite normally up to the 12-day stage, perfect histological differentiation being attained in the total absence of the peripheral sense organ. Degeneration of the differentiated cells occured later.

The general form of the body

One of the types of 'monster', the so-called omphalocephalic (Wolff, 1933a, b), has a certain intrinsic interest in relation to the epigenetic processes involved in determining the general shape of the body. It can be produced with some regularity if injuries are made (by X-rays, electro-cautery or knife) in the anterior part of the blastoderm before the appearance of the head-fold. Its morphology is best explained by the diagram (Fig. 37); the head is bent sharply downwards and protrudes ventrally, being covered with a thin layer of endoderm, while the fore-gut and pharynx develop at some distance to the anterior. The problems raised by such embryos seem to be primarily twofold. In the first place, what forces act to drive downwards the head to this abnormal degree; and, secondly, are we to conclude from them that the presumptive fore-gut endoderm can differentiate independently of any stimulus from the neural system or even from the appropriate splanchnic mesoderm? The second question has been discussed on p. 105 and it was then pointed out that the cardiac mesoderm develops in the neighbourhood of the displaced gut-endoderm, so that it is by no means impossible that the latter differentiates in response to an inductive stimulus. As regards the first, Wolff suggests that the main factor in producing the distortion is an arrest in the growth of the head, but it seems that in later stages this organ is not very much smaller than usual.

The question has also been discussed by Grünwald (1941) on the basis of a number of naturally occurring malformations. He points out that the proximate cause of omphalocephaly is a bending of the neural axis towards the ventral side at a more posterior level than this bending normally occurs. This can be brought about either by a general shortening of the head by a growth-inhibition of the kind mentioned by Wolff, or by mechanical pressure, either experimental (Grünwald, 1936 b) or consequent on the presence of a twin (Grünwald, 1937 b).

Grünwald (1941) also describes and discusses a number of other naturally occurring malformations of the head. In most of these the neural tube is flat and unfolded (so-called platyneuria). In many cases, the head-fold is at least partially missing; but the fore-gut may still be comparatively normal. There can be no doubt that these two folds are not, as is often thought, simply complementary to one another and mutually dependent. On the contrary, they seem to be largely independent of one another. This independence had already been remarked on by Waddington (1936, p. 226) and Abercrombie (1937, p. 313).

Pronounced curvatures of the body axis have also been produced by the application of more or less deleterious chemicals to embryos aged about 48 hours. Lallemand (1938, 1939a, b) found that if colchicine is applied to blastoderms of 48 hours' incubation, strong flexures occur, which may cause the vertebral column to be bent dorsally into a sharp curve like a letter J. This condition is known as strophosomy. Paff (1939), who investigated earlier stages of embryos which had been similarly treated with colchicine, describes a number of rather irregular malformations induced by the treatment, and suggests that the drug causes a stimulation of cell division as well as producing the well-known inhibition of anaphase movement. Gabriel (1946b), applying the drug by means of small impregnated pieces of agar, found that the most sensitive region lies between the omphalo-mesenteric veins and the hind-limb buds. Ancel & Lallemand (1940, 1941) and Ancel (1947) subsequently showed that the same effect can be produced by ricin and abrin. The flexure of the spine dorsally usually prevents the ventral body wall from closing, so that the intestines protrude through an extensive hernia, a condition known to teratologists as coelosomy. Ancel (1947) discusses a number of other agents which can cause such hernias, particularly unduly high incubation temperatures, generalised oedema, and localised destruction of the body wall due to haemorrhages, which may be produced by the action of drugs such as trypan blue, quinine or strychnine. Extensive teratological effects of trypan blue on mammal embryos have been reported by Gillman, Gilbert & Gillman (1948). In the chick, coelosomy may also follow failure of the amnion, which may be provoked by certain drugs, such as abrin, ricin, trypaflavine, testosterone, progesterone, desoxycorticosterone (Ancel, 1947).

This may be the most appropriate place to mention a small investigation by Waddington (1937a) on the general conformation of the head of the embryo. At the end of the second day of incubation, the embryo is normally affected by a twist round the longitudinal axis, which brings the left side of the head underneath, and also by a series of flexures in the cervical and cephalic regions, by which the whole head becomes curved into a sharp crescent, within the arms of which the heart lies. In embryos cultivated *in vitro*, the direction of twist is not infrequently reversed, so that it is the right side which lies beneath, and the cephalic crescent is open on the left side; in such cases the heart is placed on the left side instead of on the right. This suggested that the heart might be causally related to the appearance of the flexures, and this suggestion was confirmed by culturing embryos from which the heart had been removed in the 7- to 12-somite stage. No twisting round the axis appeared in these, and the greater part of the flexure was also abolished, although the sharp bend between the fore- and mid-brains (the so-called cranial flexure) developed in the normal way. There was, of course, no circulation developed in these embryos. The lumina of the arteries, particularly in the head, became extremely distended, but the embryos developed quite healthily until a stage well beyond that at which the circulation normally begins. They appear to have survived longer than those described by Cairns (1941) in which the peripheral circulation was destroyed.

Huber (1949) has recently reinvestigated the causation of the curvatures of the embryo. He finds that the cranial flexure is a direct result of local growth processes in the fore-brain and cannot be suppressed. The general curvature of the head failed to occur when the heart was removed, but he showed that the main factor concerned is not the heart itself but rather its envelopes (pericardium etc.). The amnion also plays a part in producing the normal curvature.

Mayer (1942) investigated the relation between the orientation of the embryo (i.e. the position of the head of the embryo towards or away from the observer when the blunt end of the egg is to the left) and the occurrence of situs inversus. He found no essential connection, and also observed that embryos may change their orientation considerably during incubation. Moreover, embryos which begin by turning on to their right side may later twist so as to bring their left side against the yolk, which is the most usual position.

Later stages of the eye

Good differentiation of isolated eye-cups from late somite stages have been described by Danchakoff (1924, 1926), who used the chorio-allantoic method, and Strangeways & Fell (1926), who explanted their materials into tissue cultures, and Joy (1939) using the coelom as the isolation site. Alexander (1937) states that both sensory retina and lens from rather younger donors (up to 8 somites) were considerably inhibited on the chorio-allantois and in grafts to the body wall of young host embryos. Kock (1933) found that the inhibition was more severe on the retinal component, so that some of his grafts finally contained more or less isolated lenses.

Knowledge of the epigenetic situation in the eye of the chick of post-neural-fold stages goes back to the early years of the century, when Barfurth & Dragendorff (1902) performed defect experiments on embryos in the shell. They showed that if the whole eye-cup is removed, it cannot be regenerated, but that if fragments

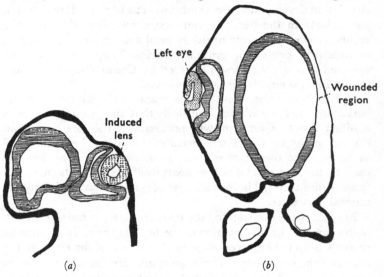

Fig. 47. *a* Lens (dashed) formed after the presumptive lens ectoderm and attached mesoderm was removed from the right optic region of a 5-somite embryo. *b* Repair of fore-brain, after removal of entire right half, with over-lying tissues, from a 15-somite embryo. Note that the ectoderm and mesoderm have grown over the wound leaving a space appropriate to the neural tissue, which, however, has not yet covered the wound. (After Waddington & Cohen.)

are left in place, these may be reorganised to develop into a complete organ. Dragendorff (1903) pointed out that this involves the production of a new lens from non-presumptive ectoderm, a phenomenon which we should now attribute to an inductive process. Similar results were obtained more recently by Reverberi (1929 *a*, *b*) and by Waddington &. Cohen (1936), the last named operating on embryos cultured *in vitro* (Fig. 47 *a*). Comparatively small fragments of eye-cup, isolated from the rest of the embryo, can also become remodelled into more or less complete eyes, although there seems some likelihood that the choroid fissure is not formed in them (cf. Waddington & Cohen, fig. 17); it is known that in the Amphibia this is determined at a much later stage than the eye-cup as a whole (cf. the review of Mangold, 1931, p. 236).

Various attempts have been made to determine whether the eye-cup can induce a lens from ectoderm of more distant parts of the body. Danchakoff (1924, 1926) and Hoadley (1926) pointed out that in chorio-allantoic grafts the morphogenesis of brain and eye is often extremely abnormal, but nevertheless lenses are usually found when the eye-cup comes into close contact with embryonic ectoderm, and suggested that this must frequently have been ectoderm other than the presumptive lens, thus indicating an inductive action. More direct methods of transplantation under the ectoderm of host embryos have been employed by Waddington & Cohen (1936), Alexander (1937) and van Deth (1939, 1940). The first named authors obtained a small number of feeble induced lenses when optic vesicles (either accompanied by or freed from their normal lens ectoderm) were grafted into streak stage blastoderms, or in one case, a blastoderm of 20 somites (Fig. 48). Alexander was more successful and obtained moderately well differentiated lenses from the host ectoderm covering transplants of optic tissue from embryos aged up to 40 somites. He claims that his results show a gradient of lens competence, the whole body being reactive up to the 4-somite stage (though better results are obtained in the anterior than the posterior parts); while, subsequent to that stage, only the head and neck region remains capable of having lenses induced in it. Van Deth has also described a small number of successful inductions performed in a similar way, and in one of these the host was of 11 somites, the graft lying in the level of the eleventh somite: thus Alexander's neat system of gradients cannot be wholly correct. In other experiments, van Deth cultivated *in vitro* isolated optic vesicles which were placed in contact with sheets of epidermis taken from various regions of the body. This technique gave rather better results and produced a fair number of successful inductions. They included one in which the lens was induced from skin taken from the dorsal part of the trunk of an embryo of 20 somites. By the third day of incubation, however, lenses can be induced only from the presumptive corneal ectoderm, although the head epidermis may show a feeble reaction of thickening.

It is seen from these results that the ectoderm from the immediate neighbourhood of the eye forms a lens more easily than ectoderm from further removed regions, and retains this capacity to a later stage. One is therefore tempted to inquire whether it can form a lens without the necessity of an inducing stimulus from the eye-cup. Waddington & Cohen removed the right optic evagination in early somite stages, before it had touched the epidermis, and reported the later development of a slight thickening

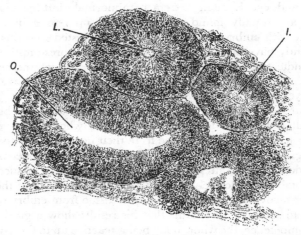

Fig. 48. The right optic vesicle of a 20-somite embryo was grafted into the area pellucida of the same embryo. The graft has formed optic cup (*O.*) and lens (*L.*); and a second lens (*l.*) is present, induced from the ectoderm, with which it is closely connected in other sections of the series. (After Waddington & Cohen.)

in place of the lens, although no lens fibres appeared (Fig. 49). They suggested that conditions were similar to those in *Bombinator pachypus* and *Amblystoma punctatum*, in which lentoids are formed in the absence of the eye-cup, and which served Spemann as models for his concept of 'doppelte Sicherung'. Van Deth was not able to confirm the existence of such thickenings, but Danchakoff (1926) claims to have found the same effect in chorioallantoic grafts and it seems unsafe to reject completely the possibility that the presumptive lens ectoderm possesses some degree of capacity for self-differentiation independent of the inducing stimulus of the optic cup.

McKeehan (1950)[1] in a short paper has described the cytological phenomena associated with the induction of the lens. He states that the nuclei in the cells of the retina are definitely arranged, approximately at right angles to the surface, before contact is made with the ectoderm of the head. As soon as this contact is

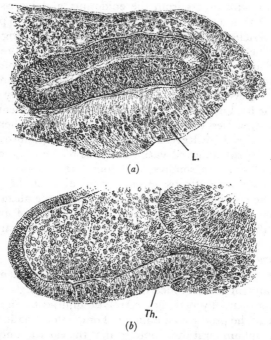

(a)

(b)

Fig. 49. The optic region of the fore-brain was removed from the right side of a 4-somite chick, the ectoderm being left intact. After 70 hours' cultivation, the head is much flattened; but on the unoperated side (a) the formation of fibres is well advanced in the lens (L.), while on the other side (b), in the absence of the optic cup, only a slight thickening has occurred (Th.). (After Waddington & Cohen.)

established, the cells of the ectoderm begin to elongate, to lose their vacuolisation and to exhibit an orientation of the nuclei. These changes extend throughout the area of contact, and are the first morphological indications of the inductive action; they appear at the same time as the basophily described by Gallera & Oprecht (1948, p. 131).

[1] See also Weiss (1950 b).

It is well known that in the Amphibia and certain other groups, a lens can be formed not only in the normal way from the superficial epidermis, but also by so-called Wolffian regeneration from the edge of the iris. Alexander (1937) observed lenses which suggested to him that they had arisen from a mingling of cells of the eye-cup with others from the epidermis, and Dorris (1938a) and van Deth found clear evidence that lenses may be produced by optic vesicles, unaccompanied by epidermis, growing in tissue culture. The appearances in the chick are slightly different in the details of histogenesis from those characteristic of Amphibia, but the phenomenon seems essentially similar. Van Deth and Dorris have also observed the origin of lens-like bodies not only from the edge of the iris but from both the retinal and the pigmented layer of the eye-cup. Less conclusive evidence of the same phenomenon has been offered by Reverberi (1930) and Dragomirov (1931). Amprino (1949) removed various regions of the optic cup from embryos of 6 to 12 somites, and found that lenses might be regenerated from optic tissue only when at least the proximal third of the cup had been left in place. The same author (1950) shows that when there is incomplete replacement of an excised eye-cup, the extrinsic muscles of the eye develop defectively.

The ear and nasal placode

Szepsenwol (1933) appears to have been the first author to advance any suggestion concerning the mechanism of development of the avian ear. By making electrolytic injuries to unincubated blastoderms, he produced a number of omphalocephalic monsters (p. 105), and amongst these noticed that the ear was only present when the associated acoustico-facialis ganglion was visible; he therefore suggested that the latter structure is the inductor of the auditory placode. At about the same time, Dalcq (1933) and Holtfreter (1933) showed that in the Amphibia this nerve is certainly not the sole or even the main inductor of the ear, which indeed begins to be visible before the ganglion is formed. Holtfreter, in particular, has shown that not only are there several tissues which co-operate to produce the normal ear, but that each tissue separately is quite able to induce an ear without the help of the others. The ear, in fact, seems to be one of the most easily induced organs in amphibian embryos.

In a special investigation, Waddington (1937b) showed that

the situation is somewhat similar in the chick, in that several factors seem to be involved; but it appears that the avian ear is less easily produced than that of the amphibia, and it seems likely that a fully normal result is not obtained unless all the factors are simultaneously co-operating.

The presumptive ectoderm of the otic placode was removed and grafted in isolation into other sites of the body, mainly among the head mesenchyme. It did not develop any sign of a placode unless taken from embryos of more than about nine pairs of somites. Not till the 21-somite stage did the isolate give an ear vesicle which appeared to be completely normal. It was also

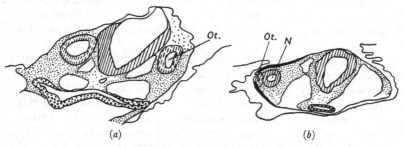

Fig. 50. *a* The presumptive auditory epithelium was removed from the right side of a 12-somite embryo; after 24 hours a small otic vesicle (*Ot.*) has been induced in the ectoderm which covers the wound. *b* The neural fold (*N.*) from the otic region of an 8-somite embryo, grafted under the ectoderm to the left of the head, has induced a small otic placode (*Ot.*). (After Waddington.)

apparent that the presumptive otic ectoderm is normally subjected to an inducing stimulus, since the nearby ectoderm which grows over the wound from which the presumptive placode has been removed is usually found to develop into an auditory rudiment. These regenerated placodes are the better developed, the earlier the time of operation; the capacity to form them seems to have lapsed completely by the 21-somite stage (Fig. 50*a*).

In order to investigate the nature of these inductive influences, the wall of the neural tube in the ear region was removed from embryos of less than nine pairs of somites; the presumptive otic ectoderm, which we have seen to be incapable of self-differentiation at this stage, was found to form a placode. It is not surprising to find that otic placodes appear in embryos in which regeneration of the brain is considerable, but they have also been found in embryos in which the injury to the neural system has been very little restored, and in particular in cases in which no acoustico-facialis ganglion has made an appearance, a result which disposes of Szepsenwol's suggestion that this is the essential otic inductor. We must conclude, in fact, that the induction of an otic placode can occur in the absence of either the neighbouring wall of the brain or of the acoustico-facialis ganglion. If, however, both the wall of the brain and the presumptive otic ectoderm are removed, no new placode is formed from the epidermis which covers the wound. This seems to indicate both that the brain exerts a certain inductive influence which can reinforce that of the other inducing tissues, and that the presumptive otic ectoderm of these early somite stages reacts to a slighter stimulus than does the nearby epidermis.

Inductive action by the wall of the neural tube was directly demonstrated by grafting it under the ectoderm of the head in embryos with the first few somites. In a few cases, small induced otic vesicles were formed (Fig. 50b). It is not known whether this reaction can be brought about in other parts of the body, or whether the competence is confined to the head. In one such case, not only was there a small induced vesicle accompanying the grafted neural tissue, but the presumptive ectoderm which had been left intact at the site from which the neural tissue was taken had also been induced, by its surroundings, to form another small otic placode. This demonstrates clearly the existence of multiple inductors in the development of the avian ear.

The differentiation of the otic vesicle from older stages has been studied in explants *in vitro* by Fell (1928), and Stcherbatov (1938); Waterman & Evans (1940) and Evans (1943) have described it in chorio-allantoic grafts. The latter author took grafts from embryos of 12–43 somites, and found that, as is the case in most such experiments, good histogenesis might be carried out by isolates from young stages, but that it is only the later rudiments which are capable of relatively normal morphogenesis in such an abnormal environment.

No special experiments have been made on the determination of the nasal placode. However, Waddington & Cohen (1936) observed that after removal of one half of the fore-brain, with the overlying ectoderm, a nasal placode might be formed on the injured side from the epidermis which grows over the wound. Since this cannot well be presumptive nasal ectoderm, one must conclude that the placode is formed as a result of an inductive stimulus. In the Amphibia, Zwilling (1934) has shown that the nasal placode can be induced by the fore-brain, and it is probably this organ which plays the major part in its induction in the chick. Street (1937) has described chorio-allantoic grafts of the nasal rudiment.

The hypophysis

The development of the hypophysis in chorio-allantoic grafts has been described in particular by Stein (1929, 1933). It was found to appear only rarely from grafts taken from just anterior to the node in head-process and head-fold stages, and Stein suggested that the histogenesis may be very dependent on the realisation of the normal anatomical relations. The truth of this has been largely substantiated by Hilleman (1943), who made localised electrolytic injuries in the region of the pituitary primordia in the 33-hour embryo. Although the results of the investigation should still be considered as of rather a preliminary nature, there was some evidence for a considerable series of inductive interactions during the development of the gland. Thus the infundibular region of the neural tube appears to induce the stomodeal ectoderm to form Rathke's pocket and the epithelial lobe, while the latter in turn influences the development of the saccus infundibuli into a pars nervosa. The inductive influence of the infundibulum seems to have already had some effect by the 33-hour stage, since a small epithelial stalk can develop from this stage in the absence of the infundibulum; on the other hand, the inductive process can continue into later periods, since if the stomodeal ectoderm is removed from the 33-hour embryo, the extraneous ectoderm which grows over to cover the wound can be induced to form a normal pituitary. This situation is exactly similar to that which we have met in connection with the lens and otic vesicle.

Fugo (1940) performed hypophysectomy of the embryo at the 12-somite stage, by removing the whole fore-brain. The primary morphogenesis of the rest of the body was carried out normally (although the missing region was not regenerated). In the second half of the incubative period, however, the embryos were considerably undersized, presumably owing to the absence of a pituitary growth hormone or hormones.

The nervous system

The causal processes operating during the later development of the nervous system of vertebrates have recently been summarised by Piatt (1948) and his review, together with that of Detwiler (1936), may be consulted for a general picture of the situation in other forms.[1] This section will deal only with those studies which have been made on the avian embryo.

Most of the students, who concerned themselves primarily with the neurological picture, have claimed that there is little capacity for regeneration in the chick neural tube (Ferret & Weber, 1904; Fugo, 1940; Rhines, 1944; Rhines & Windle, 1944). This is in some contrast to the opinions expressed in previous paragraphs (p. 146). It is, of course, possible that the apparent regeneration described, following operations on chicks *in vitro*, is only a morphological one, and that if it were possible to keep the embryos alive until the differentiation of the nervous tracts, it would be found that the defects had not been made good. However, this does not appear very likely. It seems more probable that the cases in which regeneration fails to occur are those in which a complete section of the tube has been removed, whereas in those for which regeneration is claimed, the defect only affected part of the cross-section. It also appears to be the case that regeneration is prevented when a piece of tissue is killed and left *in situ*, as in the cauterisations of Hilleman (1943) and the X-irradiations of Wolff (1936), but the latter have not been fully described from the neurological point of view.

Mosaic development, or at least a lack of mutual interaction within the length of the tube, has also been described by Rhines and Rhines & Windle, who found that the development of the cranial motor nuclei was little influenced by the presence or absence of longitudinal tracts (but see Levi-Montalcini, 1949). As far as the cord is concerned, Bueker (1943, 1945) found that the motor columns developed almost entirely normally in segments transplanted to other sites in the lateral parts of the body and therefore lacking the longitudinal tracts. Levi-Montalcini

[1] Cf. also the recent collective work, Weiss (1950).

(1945) observed the same result in segments grafted on to the chorio-allantois, and Hamburger (1946) confirmed the result by isolating sections of the cord *in situ* by pieces of tantalum foil. Both Bueker and Hamburger discuss the reasons why the earlier, contrary opinion of Williams (1931) should be rejected; and the arguments adduced by Szepsenwol (1940a) in favour of the importance of longitudinal tracts for neurone differentiation also seem insufficient.

The influence of peripheral factors on the development of the chick central nervous system has been particularly studied by Hamburger and his pupils. In 1934 he showed that extirpation of the wing bud causes considerable hypoplasia of the spinal ganglia and of the lateral motor cells. The effects persist till long after hatching and are probably permanent (Bueker, 1947). Grafting of an extra limb causes hyperplasia, which may involve an increase in motor cell numbers as high as 80 % (Hamburger, 1939a). Baumann & Landauer (1943) found that a similar situation may arise through genetically conditioned abnormalities of the periphery, since hyperplasia was observed in the motor columns supplying the legs in a polydactylous strain of birds. Tsang (1939) found the same situation in polydactylous mice, and suggested that the nervous hypertrophy was the cause of the peripheral, but both Chang (1939) and Hamburger (1939a) give reasons for rejecting this suggestion. Later work of Bueker (1943, 1945) has completely confirmed that of Hamburger on the effect of removal or transplantation of the extremities; and Szepsenwol & Goldstein (1938) and Szepsenwol (1940b, c) have demonstrated an influence of somite mesoderm on neurone differentiation and fibre growth *in vitro*.

The nature of the central changes produced by peripheral over- or under-loading have been discussed by Hamburger & Keefe (1944). They showed that the total cell count in the ventral half of the cord is unchanged, the excess or defect of motor cells being compensated by opposite changes in the numbers of non-motor cells. They therefore conclude that the peripheral factors influence the number of cells which differentiate into the motor type, rather than the total number which appear. Such an influence they suppose to be transmitted through the dendrites of the first motor cells which innervate the peripheral area. Barron (1943, 1946, 1947) has also suggested that a cell which makes contact with the periphery is thereby caused to stimulate undifferentiated cells in its neighbourhood to develop, and if such a process continued until the periphery was 'saturated', it would produce the type of hypo- or hyperplasia which is actually observed.

It is clear, however, that the concepts of hypo- and hyperplasia are too crude to serve for a full analysis of the situation. Hamburger (1948) has begun a fuller analysis of the phenomena in the cord by a detailed study of the mitotic patterns. But the point has been particularly emphasised by more recent studies on the changes produced in spinal ganglia by peripheral over- or under-loading. Levi-Montalcini & Levi (1942) drew attention to the occurrence of degenerative phenomena during the development of certain ganglia. In a detailed investigation of the lumbo-sacral ganglia (1943, 1944), they distinguished between two types of cell, a ventro-lateral group which grow and differentiate very early, and a medio-dorsal group whose differentiation and growth begins later and progresses more slowly. There is considerable evidence that the former are tactile exteroceptive neurones, and the latter proprioceptive. They behave quite differently in response to peripheral factors (Hamburger & Levi-Montalcini, 1949). The presence of an additional limb, or the absence of the normal one, leads first to changes in mitotic activity, the number of cells forming the ganglion being altered by up to 20 % either way. If an extra peripheral load is present, the majority of these additional cells begin differentiating as ventro-lateral cells, but little or no addition is made to the number which appear as medio-dorsal cells. The mechanism of this effect has been studied by Barron (1948). He shows that the initially differentiated motor neuro-blasts are at first unipolar, but that when they make contact with the periphery, they become bipolar; these bipolar cells then apparently exert an effect of an inductive character on the still indifferent cells in their neighbourhood, causing them to develop into motor neurones.

When the normal limb is removed, the numerical hypoplasia consequent on the reduced mitotic activity is exaggerated by the occurrence of degeneration. This affects primarily the ventro-lateral group, which break up and are removed by macrophages at about the time when their neurites have reached the base of the extirpated limb (5–6 days of incubation). In normal unoperated embryos a very similar process occurs in the development of the cervical and thoracic ganglia, but not in those innervating the limbs. The medio-dorsal cells are much less drastically affected, but eventually undergo a process of atrophy. A similar degeneration has been noticed in the ciliary ganglion following the extirpation of its peripheral field (Amprino, 1943).

The need to distinguish between different cell types in the response to peripheral factors is emphasised by Hamburger &

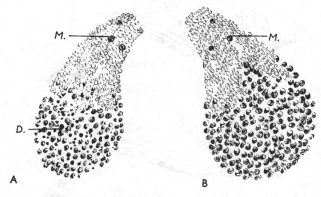

Fig. 51. The right wing bud was removed from a 6-day embryo. Degenerating cells (*D.*) can be seen in the right brachial ganglion (no. 16), A, while few, if any, are visible in the left ganglion, B; mitoses at *M.* (After Hamburger.)

Levi-Montalcini in connection with the interesting experiments of Bueker (1948), in which grafts of mouse sarcomata were made into the site of an excised limb bud or into the flank. These grafts become innervated exclusively by sensory fibres, and the spinal ganglia become hyperplastic, whereas the lateral motor columns of the cord are reduced. The hyperplasia of the ganglia shows itself, however, almost exclusively in the overall volume, the number of cells being only insignificantly greater than normal. Bueker speaks of a general increase in cell size, but Hamburger & Levi-Montalcini point out that his figures seem to show that what has happened is an increase in the proportion of the large ventro-lateral exteroceptive cells, accompanied by a reduction, presumably consequent on the absence of the limb-musculature, of both the medio-dorsal proprioceptive ganglionic cells and or the lateral motor cells of the cord.

Little study has yet been devoted to the effects of peripheral factors on the sympathetic ganglia. Simmler (1949) finds that the initial phases of differentiation of the brachial sympathetic ganglia are unaffected by wing extirpation, but in later stages a considerable hypoplasia was apparent.

The innervation of transplanted limb buds in the chick was studied by Hamburger (1939 b). As in the Amphibia, the transplant may be entered by nerves other than those which would

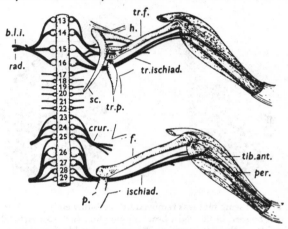

Fig. 52. The innervation of a transplanted hind limb: *b.l.i.* nervus brachialis longus inf.; *crur.* crurialis; *f.* right femur of host; *h.* humerus of host; *ischiad.* ischiadus of host; *p.* pelvis; *per.* peroneus of host; *rad.* radialis of host; *sc.* scapula of host; *tib. ant.* musculus tibialis anticus of host; *tr.f.* femur of transplant; *tr. ischiad.* ischiadus of transplant; *tr.p.* pelvic girdle of transplant. (After Hamburger.)

normally serve the limb. There is thus a non-specific attraction exerted on the nerves (cf. Hoadley, 1925 b; Hunt, 1932). In spite of its abnormal origin the pattern of the nerve within the limb may be quite typical, which must certainly indicate that it is determined by factors intrinsic to the limb rather than to the nerves (Fig. 52).

The extensive and important work of Weiss on the factors influencing nerve growth falls rather outside the scope of this book and there is no space to summarise it here (for review, see Weiss, 1941).

The somites and notochord

The initial segmentation of the somites has been discussed on p. 126. Very little is known of the causal mechanisms involved in their further differentiation. Murray (1928 b) showed from the evidence of chorio-allantoic grafts that dermis, including feather germs, may arise in grafts from regions lateral to the somites, and concluded that in normal development this tissue is not formed exclusively from the dermatome, but arises also from the lateral and ventral somatopleure. More recently, Saunders (1948 b) used carbon marks to show that the somites do not contribute either to the wing or its girdle. Rawles & Straus (1948), with a similar technique, found that the edges of the somites in the 27- to 30-somite stage correspond only to the dorsal part of the body wall. The ventral three-fifths of the ribs develop from lateral plate mesoderm, and it was shown that this material can form cartilage and muscle when isolated on the chorio-allantois.

Weber (1941) has published a remarkable account of experiments which, it is claimed, demonstrate that when an optic vesicle is transplanted to a site in the trunk region where it comes fairly close to, but not actually in contact with, the notochord, the latter organ is inhibited and fails to develop fully. Further details of the operations and more convincing photographs of the results would seem to be necessary before such an unexpected relationship can be accepted as proved.

The skeleton

An extensive series of investigations has been made by Fell (1932, 1933, 1939), Fell & Robison (1929, 1930, 1934), Fell & Canti (1934) and Jacobson & Fell (1941) on the development of the skeleton in vitro. Some of these papers have been noticed in the section on limbs (p. 173), and many of them deal with the details of histological differentiation and fall outside the field of this book. The most important of these studies, from the point of view of general developmental mechanics, is that on the sternum. (Fell, 1939). It was found that the presumptive sternal tissue forms two narrow strips of the lateral body wall, which in normal development come together and fuse in the ventral mid-line. When isolated, each rudiment develops into a remarkably perfect half-sternum, although some articular surfaces, e.g. for the coracoid, are not formed. The main outlines of the shape, however, are certainly due to intrinsic factors, although they may be modified *in vitro* as a consequence of the differential effects of the culture conditions on the rates of differentiation and of growth. Even the movement of the two rudiments towards each other, which leads to their eventual fusion, is produced by intrinsic forces and is not a mere passive reaction to the push of the elongating ribs. It can in fact be carried out in isolated explants from which the ribs are excluded. Further, if the two rudiments are rearranged so that their dorsal sides, instead of their ventral sides, are facing each other, the two half-sterna move in their original directions and are thus carried away from one another instead of towards each other. Fell suggests that the movement is primarily the result of active amoeboid migration of neighbouring undifferentiated cells, which drag the sternal rudiments after them; but the nature of the forces which determine the direction in which these amoeboid cells move remains obscure. A secondary role is played by degeneration of cells in the ventral mid-line.

Jacobson & Fell (1941) have also described the origin and development of the structures associated with the mandible, and tested the capacity for bone-formation *in vitro* of the various elements. They found that, as might be expected, the various regions were determined for specific types of development considerably before any sign of histological differentiation was apparent.

The limbs

From the time of Lillie (1904) and Peebles (1911) onwards, many workers have demonstrated the power of the limb rudiments of the chick to continue differentiation after separation from the remainder of the body. Murray & Huxley (1925), Murray (1926, 1928 b), Murray & Selby (1930 b), Selby & Murray (1928), Hunt (1932) and David (1936) used the chorio-allantoic technique, and obtained recognisable, though considerably distorted, limbs from transplants isolated in the early stages of bud formation. Murray (1926) made successful grafts from an embryo of 27 somites (48-hour stage), and concluded (1928 a) that the lack of differentiation from earlier stages was due to the unfavourable conditions rather than to an intrinsic lack of capacity in the graft. This suggestion was fully confirmed when Hamburger (1938) introduced the method of grafting fragments into the coelom of a host embryo, a situation which allows of much more normal morphogenesis, not only of the skeleton but also of the musculature (Eastlick, 1943). Hamburger obtained well-formed limbs from a high proportion of grafts from the 25-somite stage onwards, and Rudnick (1945 a) was able to observe similarly successful results from still earlier stages; the earliest blastoderm which yielded a wing was in the head-fold stage and legs were obtained from the 6-somite stage. Differentiation *in vitro* is very much less successful. Fell & Robison (1929) obtained good development of the rudiment of the femur from 5½-day embryos, and demonstrated that these produced a phosphatase in connection with hypertrophic cartilage, but 3-day limb buds gave rise only to masses of small-celled cartilage which contained no phosphatase.

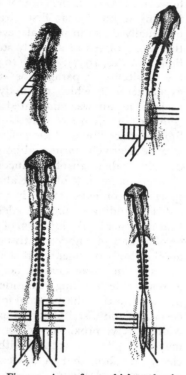

Using Hamburger's methods, Eastlick (1939c, 1941) has carried out perfectly successful heteroplastic grafts of the limbs between duck, turkey, guinea fowl and chick embryos. The foreign limb may be completely innervated by nerves from the host, and may continue development until some

Fig. 53. Areas from which coelomic grafts have yielded wing (horizontal lines) or leg (vertical lines). (After Rudnick.)

time after hatching, when immunological antagonisms begin to be effective. These vary according to the specific combinations, the duck persisting longest as a graft on the chick, a few cases lasting for some months. (Milford (1941), who grafted duck metanephroi into the coelom of chick embryos, also noted the absence of incompatibility reactions up to the time of hatching and Oakley (1938) found much the same with liver grafts.)

We have still no indication of the factors which cause the localisation of the limb rudiment in the chick, and no successful attempts have been made to induce extra limbs in a way comparable to that demonstrated by Balinsky in Amphibia. It is certain however, that, as is the case for other organs, the area capable of forming a limb is at first larger than the region which will eventually do so in normal development, and that fragments of this area isolated at an early stage can develop into a complete limb (Rawles, 1947). Wolff (1934 b, 1936) has also demonstrated the possibility of paragenesis at similar stages in the course of experiments in which the body region between the individual limbs of a pair was eliminated by X-irradiation; following this, 'symmelian' monsters may be formed, furnished with a single limb produced by the fusion of the two rudiments, and such limbs may be relatively normal, although usually exhibiting some polydactyly or other signs of their incompletely regulated double nature.

More recently, Wolff & Kahn (1947 a, b) have published two short notes on experiments in which limb rudiments were cut in half longitudinally and the edges of the wound prevented from healing either by the insertion of a fragment of vitelline membrane or of a graft of embryonic tissue (which in some cases was itself another limb rudiment). If the operation is performed in early stages (from 20–30 somites, i.e. before the limb bud has formed a noticeable swelling) considerable paragenesis occurs. This usually results in the eventual formation of a single limb in which certain of the distal elements are reduplicated, i.e. polydactyly. More rarely, proximal elements, such as the femur or tarsus, may be duplicated, with or without associated polydactyly. It is clear, therefore, that in this early stage the rudiment has still considerable plasticity.

From the earliest stage at which differentiation can be obtained, the limb rudiments appear to be definitely determined either as wing or leg, and intermediates have not been found, although of course, there are many imperfectly developed specimens which cannot be classified with certainty. Peebles (1911) had claimed that when wing and leg buds were interchanged, the grafted tissue might differentiate in accordance with its new position rather than its original fate; but Saunders (1948 a) found (in a small number of cases only) that this was not the case, the limb quality of whole buds being unchanged even when attached to the site of the limb of the opposite kind; however, fragments of bud, not containing the 'apical cap', may suffer change in their limb quality (p. 172).

Hamburger (1938) and Rudnick (1945a) also found that the dorso-ventral and antero-posterior polar axes are fixed from the earliest stages; or at least they are not altered by any influences which can be brought to bear by the surroundings of the graft on the coelom wall.

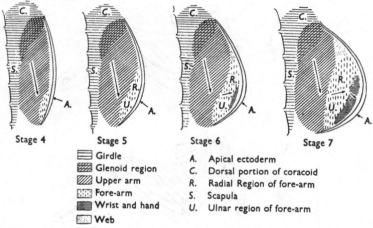

| Stage 4 | Stage 5 | Stage 6 | Stage 7 |

    Girdle   A.   Apical ectoderm
    Glenoid region   C.   Dorsal portion of coracoid
    Upper arm   R.   Radial Region of fore-arm
    Fore-arm   S.   Scapula
    Wrist and hand   U.   Ulnar region of fore-arm
    Web

Fig. 54. Presumptive areas for various tissues of the wing at four stages. The arrows indicate the direction of the future long axis. (After Saunders.)

Our knowledge of the later development of the limb bud has been considerably deepened by an important study of Saunders (1948a). This author used carbon marks to trace the presumptive areas of the various regions of the wing, and showed that the early limb bud consists almost entirely of material destined to form proximal parts (Fig. 54). The rudiments of more distal parts are laid down in sequence from an expanding region of mesenchyme which lies just below the tip of the bud, where it is covered by an apical thickening of the ectoderm. This region elongates during the early growth of the bud, apparently not so much by an unusually high mitotic rate, but rather owing to a loosening of the texture of the mass of mesenchymatous cells.[1] At the same

[1] Doubt may be expressed whether the figures and diagrams of Saunders are convincing in establishing the adequacy of this explanation to account for growth of the extent which actually occurs.

time histochemical studies by Weel (1948) showed that ascorbic acid is present around the edges of the bud, probably in the myogenic zone, while a rather wider fringe of epithelium and mesenchyme in a similar location stains deeply with vital dyes such as toluidin blue, the central chondrogenic part remaining uncoloured. But no special characteristics of the apical zone were revealed by the tests applied.

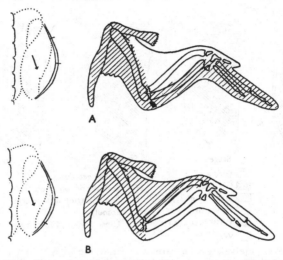

Fig. 55. The effects of excising anterior (A) or posterior (B) halves of the apical cap from wing buds in stage 5. The parts which develop after the operation are indicated by cross hatching. (After Saunders.)

Saunders also showed that when the apical ectodermal thickening is removed from a limb left *in situ*, the laying down of distal parts ceases, and a stump is developed which contains only those proximal parts corresponding to the stage to which the bud had attained at the time of the operation. Excision of the anterior or posterior half of the apical cap causes the failure of the distal parts corresponding to the half removed (Fig. 55), and leads to the formation of limbs extremely similar to those obtained by Warren (1934) following removal of the anterior or posterior half of the whole bud *in situ*. Zwilling (1949) has shown that in a genetic mutant 'wingless' race, in which the wings are greatly reduced, the apical cap of the wing bud degenerates and disappears at an early stage, after which the development of the wing appears to cease. The apical cap thus appears to play an essential role in the development of the limb, and it seems to be demonstrated that, contrary to the usual opinion, the early limb bud consists mainly of presumptive proximal material to which more distal parts will be added later. It is, perhaps, surprising in these circumstances to find Rudnick (1945 b) showing that from early isolates into the coelom the development of distal parts is better than that of proximal parts; but this indirect evidence can hardly outweigh the careful observations of Saunders.

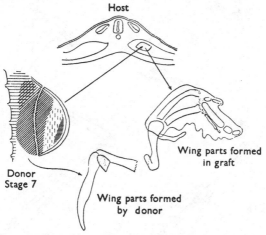

Fig. 56. Wing parts formed in a coelomic graft of the isolated tip of the wing bud, and those formed by the tissues left remaining in the stump. (After Saunders.)

In most studies which have been made on the development of isolated parts of the limb bud (such as those of Murray and his collaborators) the grafted fragment contained the apical tip, and in these the differentiations obtained seem to correspond as well as can be expected with what would be forecast on the basis of Saunders' presumptive maps (Fig. 56). Similarly, the stump from which the tip has been removed shows no sign of regulation or paragenesis (Lillie, 1904; Peebles, 1911; Spurling, 1923; Warren, 1934). In fact, the limb bud furnished with its apical cap develops as a fairly rigid mosaic, at least after the stage at which it forms a noticeable swelling, although in coelomic grafts of whole buds some slight degree of duplication may occur (Rudnick,

1945 *a*), and Wolff & Kahn (1947 *a*, *b*) observed considerably greater regulation in earlier rudiments halved along the plane transverse to the body axis (p. 168). Saunders showed, however, that regulation may be much greater in fragments separated from the apical cap. Thus if parts of the limb bud, containing both mesoderm and ectoderm but not including the cap, are removed from the dorsal surface of the wing bud and grafted into a corres-

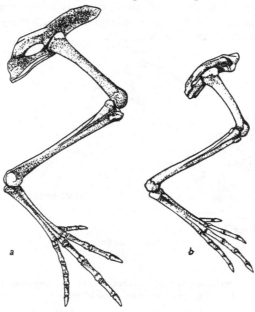

Fig. 57. Limb skeleton of a leg developed in a coelomic transplant, and badly innervated, *b*, compared with the normal leg of the host, *a*. (After Hamburger & Waugh.)

ponding place in the leg bud with reversed antero-posterior orientation, they may not only suffer a reversal of polarity, but also of limb quality, developing into parts of the leg instead of into wing. Thus the determination of limb quality and polarity, found by Hamburger (1938) and Rudnick (1945 *a*) for whole rudiments at earlier stages, must be a property of the apical region rather than intrinsic throughout the entire mass of tissue. It appears, however, to belong to the mesoderm immediately below the ectodermal cap rather than to the cap itself, since, in grafts from leg buds to wing buds, this mesoderm retains both its limb specificity and its orientation (Saunders, 1949a).

Limbs developing in atypical situations such as on the chorio-allantois, in the coelom or *in vitro* naturally lack their normal innervation. In the chick this leads to extensive atrophy and degeneration of the musculature, which proceeds further than in the frog, for example (Eastlick & Wortham, 1947). Nevertheless, the morphogenesis of the skeletal elements may be surprisingly normal, although their growth is usually somewhat reduced (about 20 %), presumably owing to the absence of some trophic action of the nerves. The degree to which the sculpturing of the

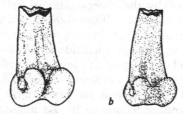

Fig. 58. The condyles at the distal end of the tibia of the right host leg (*a*) and of the coelomic graft (*b*); same specimen as in Fig. 57. (After Hamburger & Waugh.)

limb skeleton is dependent on intrinsic factors which arise within the bud itself, as opposed to extrinsic factors arising during functional activity, has been discussed in particular by Murray & Selby (1930), Murray (1936), Fell & Canti (1934), and Hamburger & Waugh (1940). As was shown by Fell & Robison (1929), extremely perfect morphological development may occur in the rudiment of the femur entirely isolated *in vitro*, but the starting point of these experiments was from a fairly late stage.

However, the results of the authors cited above leave no doubt that the shape of the various bones is to a great extent produced by internal factors. This is shown by the fact that isolated whole buds differentiate into the normal series of limb bones, and, even in the absence of any movement, may develop the usual joints (although these generally fuse at a later stage). It is also demonstrated by the fact that, if a relatively small part of the bud is isolated, the whole of it may develop into a single normally shaped bone. Since it is impossible to be certain of accurately isolating the normal bone rudiment from the mass of mesenchyme, this performance certainly provides evidence of some paragenesis, in this case a teleogenetic paragenesis; that is to say, the normal bone has been formed from tissue some of which had a different presumptive fate.

The shapes which can be produced in this way, presumably as expressions of a condition of equilibrium between forces arising in different parts of the isolated mass, are perfectly normal except for a few minor details concerned with the attachment of tendons and the passage of ligaments, for which interactions with the musculature seem to be necessary. An indication of the kind of forces which are among these intrinsic factors controlling skeletogenesis is given by the observations of Fell & Canti on the formation of the knee joint. They showed that this only develops (by 'self-differentiation') in isolates which include an adequate length of the long bones on each side of it. They suggest that the joint is produced as a response to the pressure from the two areas of chrondrification above and below it, and fails to form if, owing to the absence of one element, the pressure comes from only one side and has nothing to oppose it. As might be expected, the details of the articular surfaces were not perfect in these immobile limbs; and secondary fusion (which should be distinguished from a primary failure of fission along the joint surface) always occurred eventually. Rather similar results were found by Niven (1933) in a study of the development of the patella in the chick.

The epigenetics of bone growth and trabecular orientation have been reviewed in recent years by Murray (1936), and it is not proposed to discuss the subject here.

The tail bud

Gräper (1932, 1933) found that, when the entire tail bud is removed from embryos of 16–24 somites, the result is a complete absence of the tail. Rather similar results were obtained by Fröhlich (1926) and Zwilling (1942a), but the latter author showed that if only part of the bud is excised, the remainder may produce a tail which is normal in shape and in some cases attains the usual size; the latter result must be due to some sort of regenerative growth. If the remains of the primitive streak are removed in earlier stages (11–14 somites), regeneration can also take place, as it does in streak stages, and a complete tail may be formed. In a later paper (1946) the same author states that removal of 'the remains of the primitive streak' in embryos of 18–25 somites did not in any case interfere with tail formation. The region here concerned lies posteriorly to the zone of junction between the tail bud and the anterior differentiated tissues, which Gaertner (1949) has shown to be the zone of most rapid elongation (p. 28), and the formation of normal tails after its removal is therefore not very surprising. The excision removes, however, the whole or greater part of the first formed posterior

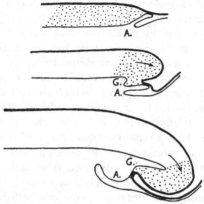

Fig. 59. Diagrammatic sagittal sections showing the early development of the allantois (*A.*) and the hind-gut (*G.*); the last section corresponds to conditions in a 30-somite embryo. (After Grünwald.)

gut diverticulum. The ultimate fate of this structure has aroused some debate, but most recent authors seem to have come to the conclusion (cf. Grünwald, 1941) that it forms the allantois (Fig. 59). This organ was lacking, or represented only by small atypical formations, in the embryos operated on by Zwilling. The caudal intestine, which develops largely at the expense of the undifferentiated tissue in the more anterior region of the tail bud, was unaffected.

The nephroi

The urinary organs of the chick first appear in the form of pronephric tubules derived from the intermediate cell mass which connects some of the anterior somites (from about the fourth to the sixteenth) with the lateral plate mesoderm in their neighbourhood. These tubules are small and often do not contain an open lumen. They are certainly not functional as excretory organs. Soon after appearing, their tips turn posteriorly, elongate, and fuse to form a continuous strand of tissue. This gradually becomes longer, stretching down to the more posterior segments, and opens to form a hollow duct, the pronephric (which is also later the mesonephric) duct. Some time after it has appeared in that region, thickenings develop in the intermediate mesoderm opposite the fourteenth to thirtieth somites and eventually differentiate into mesonephric tubules, which become active excretory organs. At a somewhat later stage, a new duct is given off from the region where the mesonephric duct joins the cloaca; this extends more or less anteriorly for some distance, and then becomes invested by the developing metanephric tubules.

We have rather little experimental evidence concerning the development of the pronephros, which is too transient, and histologically too indefinite, a structure to be recognised in chorioallantoic grafts. Waddington (1938a) has shown that if a longitudinal cut is made in a region of the embryo which is in the open neural plate stage, pronephros may develop either in connection with somite material which is isolated from lateral plate, or in connection with lateral plate which is separated from somite. It seems simplest to suppose that this represents a 'self-differentiation', the nephrogenic tissue being fairly firmly determined by that stage; but the results of Yamada (cf. Lehmann, 1945, pp. 251 *et seq.*) on the Amphibia serve to show that many other possibilities must be borne in mind. In the chick there is not yet any definite evidence that pronephros represents a certain level in a gradient whose end-points are axial mesoderm and lateral plate, as that author has demonstrated for the Amphibia.

The pronephric duct has long been considered, on the evidence of its histology, to be formed by the multiplication of the original cells derived from the tips of the pronephric tubules. The experimental evidence entirely confirms this. Boyden (1927), Grünwald (1937) and Waddington (1938a) have all shown that when a mechanical impediment is placed in the path of the elongating duct (e.g. by a transverse cut which does not heal, by cauterisation, or by a massive graft), the duct may be checked in its course, and in that case remains totally absent posterior to the blockage (Fig. 60).

The same authors all observed that when the pronephric duct is prevented from reaching the region of the mesonephros, that organ fails to appear (Fig. 61). Waddington, who investigated his embryos fairly soon after the operation, pointed out that slight condensations might be found in some cases in the mesonephrogenic mesenchyme, and this was later confirmed by Grünwald (1942). But these traces are slight, comparable at best to the lentoids which can be formed in the absence of eye-cup in certain Amphibia. It is certainly true that a properly developed mesonephros cannot be formed in the absence of a pronephric duct, and we may speak of the latter as the inducer of the mesonephros. This provides a good example of the old supposition (cf. discussion in

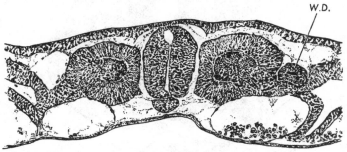

Fig. 60. A transverse cut was made posterior to the last somite of a 12-somite embryo. After 24 hours, Wolffian duct (*W.D.*) is present in the posterior part of the unoperated side, but not on the operated side (which is on left of figure). (After Waddington.)

Needham, 1931) that an organ which is vestigial in the sense that it has no function in the maintenance physiology of the animal, may nevertheless exercise an important role in morphogenesis.

Grünwald (1942, 1943) has continued the analysis of the situation, by grafting fragments of embryonic axis into the path of the pronephric duct in such a way that this is diverted and comes in contact with posterior mesenchyme which lies outside the presumptively nephrogenic region. He found that no mesonephroi were induced in this material. Thus the inductive stimulus cannot produce nephroi from any mesenchyme except that in the presumptive area or its neighbourhood. Grünwald also discovered that the presumptive tissue would react to the presence of certain abnormal inductors; in particular, neural tissue could cause the formation of mesonephric tubules, in the absence of the pronephric duct.

Boyden (1927) and Grünwald kept their embryos alive long enough to observe the effects of the operation on the metanephric duct and tubules. In the absence of the mesonephric duct, no metanephric duct is formed by the cloaca; and in the absence of this duct, not even a trace of the metanephros develops. When neural tissue is grafted into the region of the presumptive metanephros, kidney tubules are induced by it (Grünwald, 1943), but

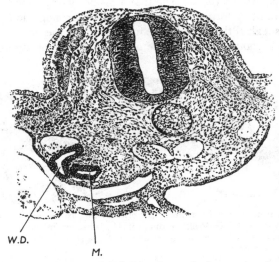

W.D.

M.

Fig. 61. A transverse cut was made posterior to the last somite of a 12-somite embryo. After two days' further incubation, Wolffian duct (*W.D.*) and mesonephros (*M.*) are present posterior to the cut on the unoperated side, but neither has developed on the operated side. (After Waddington.)

these may resemble mesonephros rather than metanephros. Boyden (1924) showed that the cloaca is, in its initial development, largely independent of the nephric ducts, although proper differentiation of its middle section (the urodeal sinus) requires contact with the Wolffian duct. In the absence of the nephric secretions, the allantois remains small.

Though the appearance of the Mullerian duct seems to be dependent on the presence of the pronephric (Wolffian) duct, the appearance of the gonads and adrenals is not thus dependent (Grünwald, 1937). The dependence of the urogenital system in its later development on hormonal influences is of course well known and it is not proposed here to summarise the voluminous and complicated literature, which belongs to endocrinology rather than experimental embryology in the usual sense.

The heart and circulatory system

The formation of twin hearts, owing to the failure of fusion of the two rudiments, was observed in early times by Gräper (1907) and in many of the experiments of Waddington (1932) and Waddington & Cohen (1936), in which operations in the mid-line had inhibited normal morphogenesis (see also Szepsenwol, 1934). It has also sometimes occurred in unoperated embryos cultivated *in vitro*; probably in these cases it is merely a mechanical effect. Nevertheless Rudnick (1935) claimed that it was produced by a differential inhibition of the tissues according as they lie near to or far from the clot, an inhibition which, however, hardly occurs unless the surface of the culture is allowed to become too dry. The separate heart vesicles begin to contract, with independent rhythms, at the appropriate time as judged by the general development of the embryo. They are roughly mirror images of each other, the left vesicle having the curvature of the normal heart while the right acquires the opposite. No proper circulation was ever observed to develop in such embryos, which therefore perished before the hearts had developed far enough for any more detailed study of their symmetry relations.

Danchakoff & Gagarin (1929) made grafts of the heart from 3- or 4-day embryos on to the chorio-allantois, and obtained good development. Kume (1935) on the other hand, who isolated hearts from embryos of 8–20 somites, found only masses of cardiac tissue, rarely with any lumen. The anterior middle region (presumptive bulbo-ventricular region) showed the least capacity to develop, the posterior middle piece (presumptive atrium) rather more. The omphalo-mesenteric (sinus) regions differed considerably on the two sides, the isolated right half showing an increasing frequency of differentiation with increasing age, while the reverse was the case with the isolated left half, which produced no heart tissue when taken from stages older than 15 somites.

Investigations on the origin and development of the heart beat have been reviewed by Patten (1949).

Until recent times very little work had been performed on the development of the embryonic blood vessels. Waddington (1937*a*) found that after removal of the heart the aortic arches develop normally for the first day or so, but that in the absence of circulation the vessels become enormously enlarged and dilated. Bremer (1928) attempted to relate the final pattern of the aortic arches to the twisting of the body and the general posterior movement of the heart during the second and third days. Wolff & Stephan (1948b) and Stephan (1949a, b) have recently published experiments in which ligatures were placed round the third to sixth aortic arches on the right side of 3-day embryos. They found that some or all of the ligated arches might fail to persist, while various compensatory phenomena occurred on the left side. Some of their results are shown in Fig. 62. The adult aortic arch may in these circumstances be formed from the fourth embryonic arch on the left instead of on the right side, or even from one of the third arches.

There has been somewhat more work on the extraembryonic circulation, which is more accessible. Grodziński (1934) employed vital marks to study the expansion of the area vasculosa and showed that the mesodermal sheet which contains the vessels spreads out between the ectoderm and endoderm even in blastoderms in which the embryonic body and blood circulation has been suppressed by temporary under-cooling. The spreading is due mainly to the high growth rate of the peripheral region, which grows much more rapidly than either the more proximal regions or the ectodermal and endodermal membranes of the area vitellina. The fact that the yolk-sac is largely independent of the embryo, and continues to expand after the death of the latter, had previously been noted by several authors (Byerly, 1926, 1932; Aliberti, 1933; Remotti 1927, 1930, 1931, 1933 a, b, c) whose papers should be consulted for further details. The amnion is also largely independent of the development of the embryo (Wolff, 1932).

The vessels of the area vasculosa, like those of the embryo itself, appear gradually by the consolidation of certain channels out of an original close network. The process has been studied in detail by Hughes (1935), who finds that there is some 'self-differentiation', in that the path of the major vessels can be recognised by the presence of rather larger vessels some hours before the flow begins. The pattern, however, is highly modifiable in response to flow, and varies considerably in normal development: thus Grodziński (1935) showed that the final anterior vein is sometimes formed from the original right vein, sometimes from the left. Wolff & Stephan (1948 *a, b*) have studied the deformations of the circulatory systems produced by introducing obstacles in the path of the expanding vessels or by eliminating

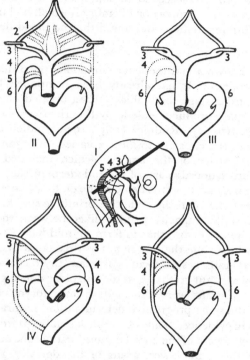

Fig. 62. I. The thread in place preparatory to ligaturing the aortic arches on the right side of a 3-day embryo. II. Ventral view of the normal aortic arches during the second half of incubation. III. Suppression of normal aortic arch. IV. Suppression of fourth right aortic arch, and compensatory formation of a fourth left arch. V. Suppression of all right arches, and formation of left third arch. (After Stephan.)

regions of the blastoderm by X-irradiation. The influence of the flow on the diameter of the vessels has been minutely studied by Hughes (1937).

At an earlier date, there was considerable controversy as to the source from which the blood is formed, some maintaining that the angioblasts arise from mesoderm, others deriving them from endoderm (for references, see Grodziński, 1934 b). The former view is now generally accepted and is certainly correct; thus Waddington (1932) found that blood develops in mesectoderms isolated from the hypoblast, but not in the hypoblasts, while Grodziński (1933) succeeded in suppressing all mesoderm formation in a certain proportion of embryos which had been undercooled before incubation, and found that no blood developed.

Endodermal organs

Little is known about the determination of the different organs of the endoderm. It has already been mentioned that in the primitive streak stage the epiblast, when isolated on the chorioallantois, may produce endodermal derivatives such as gut, thyroid, etc. (p. 92). Rudnick (1933, 1935) studied the gradual restriction of the capacity to form a variety of organs by taking longitudinal strips of the embryo which included either the median (future dorsal) parts of the endoderm or the more lateral (future ventral) parts, grafts being made on to the chorio-allantois from stages older than the definitive primitive streak. The dorsal material lost the capacity to form liver between the head-fold and 3-somite stage, and that to form thyroid before the 6-somite stage, at which time the power to form pancreas first appeared. The ventral grafts, besides forming liver and thyroid, also developed respiratory and gut epithelia, as did the older dorsal grafts.

A series of grafts were also made from embryos cultivated *in vitro*, and the same progressive determination, naturally enough, was again in evidence. Some of Rudnick's embryos were inhibited in the development of the dorsal organs (neural tube and somites); in two of these, moreover, grafts of the dorsal region yielded ventral endodermal tissues. The period of cultivation had been long enough for the embryo to have passed the stage when this capacity is lost, and the development of the ventral organs, such as the heart, had actually progressed past the corresponding stage; but the state of determination of the presumptive dorsal endoderm seems rather to have kept pace with that of the inhibited neural tube than to have continued at the normal rate. The figures are extremely small, but the result is probably what one would expect if the determination of the regions of the endoderm is largely dependent on influences from the overlying mesoderm and neural system.

Studies on the development of other endodermal organs *in vitro* or in chorio-allantoic grafts (e.g. thyroid by Rudnick, 1932, and Carpenter, 1942, liver by Willier & Rawles, 1931 *a*, intestine by Rudnick & Rawles, 1937) have also shown that the typical histogenesis, or even organogenesis may be carried out by fragments from quite early stages which contain mesoderm (see p. 87). Many other investigators, particularly in the years between 1925 and 1930, have studied grafts taken from older stages. The experiments, however, yielded little information about the progress of determination within the organs, except in the case of the lung. Rudnick (1933) found that complete histogenesis could occur in grafts from the pre-somite stage, but that the morphological development of the bronchial tree did not occur in grafts taken before the appearance of the lung bud. It is not certain whether much significance should be attached to this, in view of the abnormal mechanical conditions of the grafts on the membrane.

CHAPTER IX

THE PLUMAGE

Melanophores

In recent years, a very considerable amount of work has been done, mainly by American authors, on the development of pigment-forming cells in the chick and other birds. The literature has recently been summarised (Rawles 1948; Du Shane, 1944, 1948), and for this reason, and because it lies rather peripherally to the main theme of this book, it will here be dealt with in outline only.

The origin of melanophores from the neural crest in amphibians had been demonstrated more or less conclusively by a number of authors before 1939, but the first considerable studies on the subject were those of Raven (1933), Du Shane (1935) and Twitty (1936). These authors showed that if the neural crest is removed from a certain length of the neural plate, no pigment cells appear in that region, at least at first. It thus seemed that all the pigment-forming cells were derived from the crest. More recent work (Niu, 1947; Twitty, 1949) indicates that, in urodeles at least, pigment cells appear, after some delay, in the head and tail regions of embryos from which the whole crest has been removed; they probably come from the brain and from the tissue derived from the posterior part of the neural plate, and thus do not differ profoundly from the normal crest-melanophores in their provenance.

Very soon after this work by Du Shane, evidence for a similar state of affairs was found in the chick. Dorris (1936, 1938*b*) showed that fragments which contain part of the neural crest form typical pigment cells when grown in tissue culture; while if such isolates are grafted on to the leg bud of 3-day-old host embryos, they migrate into the host tissues and cause the formation of donor-coloured feathers (Dorris, 1939; Eastlick, 1939*b*). The converse experiment, of removing the whole crest from an embryo, is not possible in the chick, but a study can be made of the pigment-forming capacities of isolated fragments taken from various body regions at some distance from the crest. In one of the earliest studies of pigment formation in tissue culture, Koller (1929) found that mesenchyme of those breeds which form black pigment in the skin possessed the power to do so after isolation; but these isolates were taken from fairly old stages (72-hour). When isolates are made at sufficiently early stages, pigment is formed only from those which contain crest or presumptive crest (Eastlick, 1939a; Ris, 1941).

By making a very large series of transplantations of fragments to the base of the wing bud of White Leghorn hosts, Fox (1940) has been able to follow the migrations of the melanoblasts from the crest into the lateral regions of the body. It was found that these cells begin to leave the mid-line of the embryo in the region of the mesencephalon at about the 8-somite stage. The region of emigration gradually spreads posteriorly, its posterior border being always slightly anterior to the last-formed somite (by about the length of five or six pairs of somites). At first the cells move laterally only far enough to arrive at the edge of the row of somites. Fox claims that it is first in the limb regions that this limit is overstepped, since melanoblasts are found in the bases of the limb buds at about the 44–45 somite stage though they are not demonstrable in the intervening region till rather later; the data on which this conclusion is based, however, do not appear to have been subjected to any rigorous statistical evaluation, and to all appearances they do not look adequate to prove that the suggested difference is a real one. The melanoblasts could also be shown to enter mesodermal structures, such as the linings of the coelom, the nephroi, gonads, etc., where, according to Fox's data, they arrive at a rather later period than they reach the immediately overlying epidermis. In most breeds no pigment is found in these structures in the adult; but it is not certain whether the melanoblasts in the organs degenerate and disappear entirely or whether they merely fail to form pigment. Saunders (1949b) has recently shown that the migrating melanoblasts can be vitally stained with nile blue or neutral red, and the course of the migration may in the near future be more fully explored by the use of this fact.

It is thus clear that at least the greater part of the pigment cells in birds originate from the neural crest, and that most regions of the body are unable to produce pigment if neural crest cells are excluded. It remains possible that some other regions may be able to give rise to melanoblasts, as in the Amphibia, but there is no definite evidence for such an occurrence, at least in stages later than that at which the crest becomes definitely organised. In stages developmentally less advanced, the potential sources of melanoblasts are less restricted. Thus Eastlick (1940) and Rawles (1940) found that in early stages, the whole region around the node, which is capable of forming neural tissue in chorio-allantoic grafts, is also capable of giving rise to pigment cells, and Fox (1940) showed that isolated fragments of the tail bud can do the same. These phenomena presumably depend on regulation within the still labile node tissues.

It should be pointed out that other types of pigment occur in the fowl besides the melanins characteristic of the cells which are here under discussion, and that these pigments may be formed in cells which are not connected with the neural crest in any way. Thus the retinal pigment of the eye arises *in situ* (Ris, 1941), although that of the choroid and iris is of the normal melanophore type derived from the crest. The lipochrome pigments which may be found in feathers in some breeds have also no essential connection with melanoblasts (cf. Volker, 1944).

The causal factors responsible for melanoblast migration are under active study by American authors, but are still very imperfectly understood. The extensiveness of the migratory movements was very impressively demonstrated in experiments by Willier & Rawles (1940) and Rawles (1939), in which small fragments placed at the base of the wing were found to cause the production of donor-type pigment granules over the whole wing area, even in intergeneric combinations. Working with urodeles, Twitty (cf. 1949) presented evidence suggesting that the main motive force which brought about this migration was a repulsion between neighbouring melanoblasts, which caused them to scatter as widely as other factors permit. There is as yet no positive evidence for this in birds. On the other hand, something is known about certain of the 'other factors' mentioned above.

In the first place, migratory activity is much greater in young melanoblasts than in those of later stages; in fact, it is likely that the avian melanoblast is incapable of any appreciable migration after it has once formed pigment (Hamilton, 1940), although this is not true of the comparable amphibian cell. Migration is also more rapid into comparatively young hosts: by the stage of the 4–5-day embryo, whose limb buds have already a population of their own melanoblasts, grafted neural crest cells are able to make little progress in penetrating host tissues (Willier & Rawles, 1940). This may, perhaps, be due to repulsive forces of the kind postulated by Twitty, and have nothing to do with any special characteristics of the non-melanophoric host tissue. In fact, Rawles (1944) showed that even after hatching, melanoblasts from the host could migrate into grafts of similar age which lacked pigment cells of their own, having been originally isolated from the donor before any melanoblasts had reached them and cultivated as coelomic grafts where they were inaccessible to such migrating cells. Willier & Rawles (1940) also found that melanoblasts of certain breeds (White Leghorn, containing dominant white) migrate less readily than do cells of comparable age taken from breeds containing the recessive white factor (e.g. Wyandotte, Plymouth Rock, etc.).

As to the actual pathways of migration, it is probably significant that the melanoblasts tend to be concentrated along distinct surfaces, such as those of the blood vessels, nerves, etc. As Fox (1949) points out, this suggests that an important part in migration is played by factors of surface affinity and substrate orientation, of the kind which Weiss (1941, 1945) has discussed in relation to other migrating cells, such as fibroblasts, Schwann cells and nerve fibres.

Gillingham & Medawar (see Medawar, 1947) have presented evidence that in post-embryonic stages of mammals the capacity to form pigment may be transmitted from one melanoblast to a contiguous one by a process of 'infective transformation'. There is as yet no evidence for a similar process in birds, and the fact that, when a graft causes the production of pigment in a host, the resulting pigment is exactly similar in type to that proper to the graft makes it very unlikely that any process of the evocation of pigment within the host cells has played a role. On the other hand, the evidence that migration of melanoblasts really occurs during early development seems incontestable, and this suffices to explain all the phenomena known in birds, even if it is not adequate alone in relation to mammals.

The chemical constitution of melanins is not understood, and the nature of the reactions by which they are produced in the melanophores is still obscure. Present knowledge on the subject is reviewed by Lerner & Fitzpatrick (1950). It is well known that the reaction is strongly affected by the hormone balance of the organism, the gonads and the thyroid being particularly important (Willier, 1942). Several different types of melanin pigment may be formed, reds and yellows as well as blacks (cf. Mason, 1948). Recent papers on this subject, in which considerable bibliographies are given, are those of Blivaiss (1947), Markert (1948) and Trinkaus (1948).

Feathers

The avian feather is an organ which begins to develop rather late in embryonic life and is, therefore, rather on the edge of the field covered in this book. However, because the feather germ retains into the adult stage its capacity to develop new organs after the previous ones have been plucked, and also because of the striking and clear-cut form and pattern of the plumage, a very great deal of experimental work has been done on this epigenetic system, and it would seem a pity to omit this completely. The account which follows deals only with some of the points which seem of most general importance, and no attempt will be made to discuss all the details which are known.

It will be logical to begin by a consideration of the determination of feather rudiment. A general review of the literature up to 1941 has been given by Lillie (1942) but some very important additions have been made to our knowledge since then, particularly in the various papers of Lillie & Wang, which are referred to later.

Feathers grow from papillae, which appear in the chick embryo at about the eighth day of incubation. They are arranged in definite rows, which can themselves be assigned to a number of different tracts characterising the various regions of the body (cf. Holmes, 1935). The whole of a tract has a certain developmental and physiological unity and the detailed characteristics of the individual feathers (growth rate, maximum size, etc.) vary in an orderly manner throughout the whole area. These 'field' characteristics of the tracts have been described in detail by Fraps & Juhn (1936 a, b). Each feather papilla persists throughout life, and gives rise to a series of generations of feathers, the earliest members of which (the down feathers) differ markedly from the later ones. The development of the down feathers has been minutely described by Watterson (1942) but we shall not here be further concerned with them. The adult feathers not only differ

from the down feathers, but also differ among themselves, each
tract having certain specific characteristics of form and of colour
pattern (cf. Hosker, 1936).

At the time when the definitive feather is formed, the feather
germ or follicle consists of a slight hillock on the skin, from the
centre of which a canal extends downwards; at the base of this is
a more or less conical or thimble-shaped papilla, and it is from this
that the actual feather arises. The papilla is a double structure,

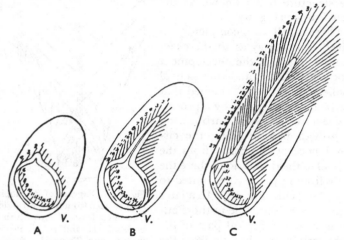

Fig. 63. Diagrams showing the development of the feather, as envisaged by
Lillie and his school (viewed obliquely from below the dorsal side of the papilla).
After the initial complement of ridges (seen in A), the new ones appear at the
ventral side of the collar *V*., and move tangentially towards the outgrowing
rachis (B and C). (After Lillie & Juhn.)

with an epidermal (ectodermal) cap fitting above a dermal
(mesodermal) core. Full accounts of the morphology have been
given recently not only by Lillie (1942) but by 'Espinasse (1939).
The first important point to notice is that the actual feather is
derived mainly from the ectodermal component. At the base of
the follicle, this epidermal cap becomes a thick ring of tissue, with
a narrow lumen which leaves room only for a thin stalk connecting
the dermal core with the deeper lying mesoderm. This thickened
ring is known as the collar. From it arise the rachis and barbs of
the feather, the mesoderm contributing only the pulp which fills
the base of the rachis. The follicles always lie at an angle to the

surface of the skin, so that the feathers will point in some particular direction, usually more or less posteriorly to the animal as a whole. One side of the follicle is, therefore, nearest the skin. This is known as the dorsal side, and the rachis is always formed in this position, the barbs being formed in the lateral walls on each side. At the diametrically opposite, ventral, side of the collar, a small secondary rachis, accompanied by a few barbs, may appear, the structure being known as the after-feather (Fig. 64).

There is still an incompletely resolved controversy as to the exact manner in which the development proceeds. Lillie and his school (Lillie & Juhn, 1932) claimed that the rachis was formed by a process of concrescence, each barb during its growth passing along the circumference of the collar from the ventral to the dorsal side, where the barbs from opposite sides united to form the rachis (Fig. 63). In a later paper (1938) the same authors admitted that the central part of the rachis originated directly from the dorsal material, but still argued that its lateral faces were formed by concrescence in this way: Lillie

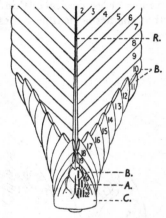

Fig. 64. Diagram of developing feather, seen from the ventral side. The 'after-feather' *A.*, with three barbs, I, II and III, is nearest the observer. The rachis *R.* arises from the far side of the collar *C.*, and bears the barbs *B.* (After 'Espinasse.)

& Wang (1941) claimed to find further evidence of this in the fact that after partial maceration in alkali, the barbs separated from the rachis bringing with them a small section of the lateral face of the rachis which they spoke of as the barb petiole. 'Espinasse (1939), on the other hand, called in question the existence of any concrescence at all, arguing that growth is at all times strictly along the length of the follicle and thus perpendicular to the collar. He was probably successful in showing that the arguments for the concrescence theory are not absolutely compelling, but it is doubtful whether the same should not be said about his own views. The matter requires investigation, possibly with the aid of vitally stained marks.

After Danforth had made many earlier transplantations of follicle-containing skin (cf. Danforth, 1929; Danforth & Foster, 1929), Lillie & Wang (1941) found that it was possible to carry out surgical operations and transplantations with the developing follicles of regenerating feathers. They first showed that if a papilla is completely extirpated, the follicle cannot regenerate a new one, and forms no further feathers. If the whole papilla is transplanted into another follicle (from which the original papilla has usually been removed), it forms a feather whose orientation and tract-specific type are those of the transplanted papilla. When incomplete parts of the papilla are transplanted, the dorsal region alone is capable of forming a complete feather with a rachis: the isolated ventral half gives rise merely to a tuft of barbs with no rachis, a result which speaks rather strongly against the concrescence theory, at least in its original form. Lateral halves, which contain an appreciable amount of the dorsal region, can form complete, bilaterally symmetrical feathers, presumably owing to regulative (telogenetic, paragenetic) processes. Similar telogenesis, however, does not occur so completely along the apical-basal axis: if the tip of the papilla is removed, the apex of the feather, although not amputated, is present in an abnormal form.

When the isolate contains only a small section of the dorsal material, telogenesis is incomplete and the feather formed is asymmetrical or possesses a barb only on one side (Lillie & Wang, 1944). Sections of either the dorsal or ventral sides of the papilla may be inserted into other papillae which have been split and opened along a radius. A dorsal sector, whose size is such that it subtends about 60° at the centre, develops into a feather with rachis wherever it is placed around the circumference of the host-papilla. This secondary feather is often quite as large as the host-feather, and it seems certain that it contains host material, which has thus been caused to enter the epigenetic system of the graft. Grafted ventral sectors, by contrast, do not form raches. If placed in the dorsal region of the host, they may split the host system and cause the formation of double feathers, each of which arises from one of the lateral faces of the host papilla: similarly, if placed ventrally, they cause an increase in size of the after-feathers. It is extremely interesting to note that the same result is obtained if a ventral graft is placed to one side of the dorsal region of the host; the effect can only be due to a weakening of this region in some manner akin to a dilution, and demonstrates that the size of the after-feather is inversely determined by the strength of the dorsal influence. The difference between the dorsal and ventral sides of the papilla must thus be, in the main, a quantitative one.

In some of these experiments, grafts from breast feather papillae were combined with saddle feather hosts or vice versa. There was evidence that the breast feather dorsal region is, in some sense, stronger than that of the saddle feather, since a breast dorsal graft into the saddle ventral region forms a larger feather than the reverse combination (Fig. 65). A more complete study of such interactions was made by Lillie & Wang (1943), in which halves of papillae from these two regions were combined. It was found that chimerical feathers developed, in which certain portions could be recognised as consisting of saddle-type barbs, others of breast-type. The most interesting combinations were those of breast-dorsal with saddle-ventral or vice versa. In the former, a longitudinal chimera was produced in which the tip of each barb was of saddle type, while the base was of breast type. In the latter, the result was quite different, the chimera being transverse, with a saddle apex and a breast base.

Lillie & Wang interpret the first of these results as providing crucial evidence in support of the concrescence theory. They suppose that each barb begins forming at its tip in the ventral (saddle) region but that, as it elongates, the base becomes contributed by ever more dorsal regions, and thus is eventually of breast type. This would certainly seem the obvious explanation, but it appears to the present author to speak directly against the concrescence hypothesis rather than for it. The concrescence hypothesis surely implies that cells which originally lay on the ventral side (i.e. in these experiments, saddle-type cells) will eventually come to be on the dorsal side next to the rachis. This is exactly what does not happen in this combination, which in this respect appears to strengthen the case for 'Espinasse's suggestion.

The whole situation, however, is not so simple, because in the combination of a saddle-dorsal half with a breast ventral the chimera produced is of a transverse instead of a longitudinal type. In this case the apex of the feather is saddle-type and there follows an intermediate zone in which the barbs have breast-type tips of

increasing length, until the whole barb in the basal part of the
feather is of breast type. Lillie & Wang suppose that this indicates
a gradual expansion of the breast material from the ventral side
until it completely replaces the saddle-dorsal tissue. This might be
taken to represent a real concrescence. The explanation, however,
is made somewhat uncertain by the fact that if such a feather is
plucked, it regenerates in the same form as before. Thus the
replacement of saddle tissue by breast tissue certainly does not
apply to the papilla as a whole. Lillie & Wang state that 'there
can be no doubt, therefore, that the composition of the *papilla*
established after healing is quite stable as regards both epidermal
and dermal constituents.... It contrasts with the instability of the
collar in saddle-breast combinations, which is a function of development and not of the organisation of the germ.' What the last
clause of this sentence is intended to convey must be left to the
intuition of the reader. It seems to the present author that a
convincing explanation of these phenomena has not yet been
provided. Perhaps one may emerge from a further study of the
mutual influence of breast and saddle tissues on each other's
growth rates, a matter which has been opened up by Wang (1945).

A further important step in the analysis was taken by Wang
(1943) who discovered a technique which made it possible to strip
the epidermal component away from the mesodermal core (by
a short treatment with 25 % salicylic acid in collodion). The
epidermis was killed, but the isolated dermal part of the papilla
could be transplanted into follicles from which their own papilla
had been removed. Such follicles are not able to form feathers,
either if left alone, or if extra-follicular epidermis is grafted into
them. When the dermal part of the papilla alone is grafted into an
empty follicle, feathers form in a considerable percentage of cases.
Thus this component acts as an inductor for the epidermally
derived feather. Moreover, it determines the orientation of the
feather, as was shown by grafts which had been rotated before
implantation. However, the tract specificity was that of the host,
saddle dermal cores producing breast feathers when transplanted
to breast follicles and so on; this is in complete contrast to the
donor-wise development found when entire papillae are grafted.
Thus the factors determining tract specificity must be located in
the follicle epidermis which grows over and covers the transplanted
dermal core.

The whole series of experiments of Lillie & Wang have produced a large body of results and are a beautiful technical achievement. It is unfortunate that nevertheless so many points, particularly the concrescence theory, remain obscure. It is also necessary to make some criticism of the terms in which they have been discussed. Lillie (1942) and Lillie & Wang (1941, 1944) speak of

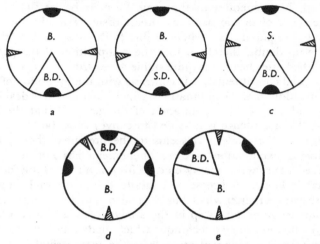

Fig. 65. Diagrams showing the results of grafting dorsal sectors of feather papillae into various positions of host papillae. The black semicircles indicate the raches of the feathers which develop, and the shaded triangles show the margins up to which each feather extends. In *a*, a dorsal sector of a breast feather *B.D.* grafted into the ventral side of a breast papilla *B.* produces a secondary feather as large as the host; in *b*, a similar graft from a saddle papilla *S.D.* produces a smaller secondary; in *c*, a breast graft into a saddle host papilla *S.* causes a larger feather to develop; in *d*, a breast graft into the dorsal side develops into a small feather while the host papilla produces two feathers; *e* shows the effect of an eccentrically placed graft in shifting the position of the host rachis. (After Lillie & Wang.)

the dorsal sectors of papillae which they transplanted into other split papillae as 'inductors' of feathers. Wang (1943) also uses the same term of the dermal component. To the latter usage, no objection can be taken, and the action can be legitimately compared to that between axial mesoderm and ectoderm at the primitive streak stage of the chick or the gastrula stage of the newt. However, Lillie & Wang, in some places at least, use the same term for the former process, in which a grafted part of the feather rudiment merely completes itself at the expense of the surrounding tissue. Although such a process is perhaps also in some sense 'an induction', it is one in which what is induced is essentially similar to that which induces. It is comparable rather to the processes which go on in combinations of echinoderm blastomeres, as described by Horstadius, than to the classical induction phenomena of the Spemann organiser. It is, in fact, an example of what has been described above (p. 113) by the term 'individuation'. The reacting host material becomes completely assimilated and takes part in building up the complete individuated organ together with the 'inducing' graft. In the more classical and properly so-called type of induction, seen in the Spemannian organiser or the reaction between epidermis and the dermal core of the follicle, many characteristics of the resulting organ (such as tract specificity) remain determined by the reacting tissue. The distinction between these two types of behaviour should not be lost sight of.

Lillie & Wang (1943) have offered an explanation of very similar kind to that used by Horstadius, namely, one in terms of gradients. They suppose that some substance or activity is present in a stronger form in the dorsal region of the papilla, decreasing in intensity along the lateral faces towards the ventral side, and they suggest that all the parageneses found can be easily understood in terms of the reconstitution of this gradient system.

The determination of tract specificity by the reacting epidermis raises a fundamental problem of the mechanism of development. In previous examples of similar phenomena, the specificity of the reacting tissue has usually been genetic in nature; for instance, one knows that in grafts between species or genera of Amphibia, the induced tissues develop into organs characteristic of their own species and not of the inducer. Such specificities can plausibly be attributed to the genes. The feather-tract specificities, however, certainly do not depend on differences in the original genotype of the zygotes involved. Must one then suppose that the course of embryonic differentiation which has led breast epidermis to have certain inherent differences from saddle epidermis has depended on some sort of controlled mutation, so that cells of the two tissues actually contain different genes? Some authors, particularly perhaps those who write on the subject of cancer, have argued in this sense (for a recent statement of the view, see Rhoads, 1949). But the evidence certainly does not compel us to that conclusion;

and if 'gene-mutation' is to be invoked to explain differentiation it would have to be of a character so special, in its regularity and orderly succession of steps, that it is doubtful if any good purpose would be served by using for it a term likely to identify it with the normal process which gives rise to the mutations detected by breeding experiments. A somewhat fuller discussion of these differences has been given elsewhere (Waddington, 1948) and in that paper other hypotheses are discussed as to the manner in which different tissues of the embryo come to develop a restricted repertoire of competences, one or other of which can be realised under the stimulus of the appropriate inductor, while no developmental performance outside the repertoire remains possible. The feather experiments are a demonstration that the specificity of competence may be sufficiently detailed to include the possibility of one tract-type of feather while excluding another.

Feather patterns

The epigenetics of the colour patterns of feathers has been extensively investigated from two rather different points of view. The earlier workers, such as Lillie & Juhn, 'Espinasse and others, were largely interested in the patterns which can be artificially produced as responses to hormone injections, using these as indicators of the underlying physiological processes going on in the feather. More recently, the methods of transplanting melanoblasts, largely due to Willier & Rawles, have made possible investigation of the interaction between pigment cells of various different genotypes and the other components of the feather germ. There is not space here to discuss either type of work in detail, and reference must be made to the reviews of 'Espinasse (1939), Lillie (1942), Rawles (1948) and Willier (1948). Only a few salient points are mentioned below.

In 1930 Juhn & Gustavson published evidence that if female hormone (oestrone) is injected in Brown Leghorn capons, a region of female pigmentation (described as salmon-coloured) will be developed in the dark brown growing feathers. They showed that the different feather-tracts of the capon have different thresholds of reaction, showing no response to injections which do not raise the oestrone level above the necessary minimum. These threshold levels corresponded with the characteristic growth rate of the tract, being lower for tracts with slow-growing feathers. Further, the different tracts showed differences in the time interval which occurs between the time of injection and the time at which the reaction becomes manifest.

Lillie & Juhn (1932) then proceeded to extend these conceptions to explain the shape of the feminised region which can be produced on an individual feather as a response to a single hormone injection or series of such injections. They showed that the threshold of response is lowest on the dorsal side of the papilla, next to the rachis, and rises ventrally. The lag between injection and the occurrence of the change varies in the opposite direction, being least ventrally and greatest dorsally. In these terms it is easy to explain the three main types of pattern which are obtained. These are

(1) A median spot near the rachis, obtained as a response to an injection which does not raise the hormone high enough to pass the threshold of the more ventral tips of the barbs;

(2) A more or less transverse band, following an injection large enough to affect the tips of the barbs; and

(3) two marginal spots, which is the response to an injection which is massive enough to pass the ventral threshold (and, therefore, *ex hypothesi* the lower dorsal threshold) but which gets excreted before the expiry of the long lag-period of the dorsal region.

This theory is a very neat example of the way in which the formation of morphological patterns can be explained in terms of variables (thresholds, time lag, etc.) which do not themselves possess the same morphological properties. As such, it has considerable heuristic value. How far it also possesses the advantage of being true is not quite so certain. In its first expression at least, Lillie & Juhn suggested that the variation in threshold and time-lag within the feather depend in the same way on the growth rate of the parts as they do in the comparisons between different feather tracts. Such dependence on growth rate has been strongly criticised by 'Espinasse (1939), and is now only held in a modified form, if at all, by Lillie (cf. 1942).

The subject developed a highly complicated special terminology, largely at the hands of Juhn & Fraps (1936). This has been critically reviewed by 'Espinasse in the paper cited above, but the whole subject still remains in some obscurity and it is not easy to decide how much is left standing of the neat theoretical picture which Lillie & Juhn originally proposed.

The pigmentation of the feathers is, of course, due to the melanoblasts whose migrations were discussed in a previous section. It remains to say something of the relations between these cells and the feather follicles. In the resting stage, when the feather is fully grown, the feather papilla appears to contain no melanoblasts. If the feather is plucked and regeneration starts, the melanoblasts migrate from the nearby dermis through the dermal core of the papilla and thus reach the epidermal collar from which the feather is growing (Foulks, 1943). During this migration they are still uncoloured and their presence or absence is best tested by grafting the tissue under investigation into the limb bud of a young embryo, when pigment will eventually be found if any melanoblasts are present.

The absence of melanoblasts from the resting papilla makes it fairly simple to arrange that a feather follicle of one kind will become populated by melanoblasts of some different genetic type, and numerous experiments have been made in which this end has been attained in various ways, for instance, by grafting melanoblasts into limb buds before the pigment cells proper to the host have reached them (Dorris, 1939; Rawles, 1939) or in the more roundabout way indicated in Fig. 66. A general review of such experiments is given by Rawles (1948). The result has always been to show that, in its main characteristics at least, the pigmentation of the resulting feather is that characteristic of the donor from which the melanoblasts were taken. The papilla determines only the tract-specificities in pigment; the general nature of the pigmentation is determined wholly by the melanophores. If the donor is of a race in which there is a sexual dimorphism in the plumage, the grafted melanophores still develop into the pigment-type proper to their own genotypic sex, whatever may be the sex of the follicle which plays host to them (Willier & Rawles, 1944).

Thus if melanoblasts from a barred race are grafted so as to populate feather papillae of a white animal, the feathers receiving them will be barred, but the exact nature of the barring will show the characteristics of the tract to which the host feathers belong independently of the region of the embryo from which the melanoblasts were taken (Rawles, 1944). When different breeds of barred fowl are used as sources of the transplanted pigment cells, the particular characteristics of these breeds (e.g. length of the periodicity) are shown by the graft-populated feathers (Nickerson, 1944), a result which demonstrates fairly conclusively that the barring is inherent in the melanophores themselves and is not imposed on them by the other constituents of the papilla.

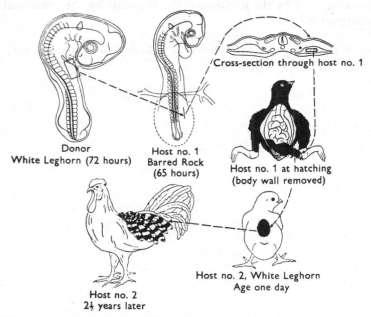

Cross-section through host no. 1

Donor
White Leghorn (72 hours)

Host no. 1
Barred Rock
(65 hours)

Host no. 1 at hatching
(body wall removed)

Host no. 2, White Leghorn
Age one day

Host no. 2
2½ years later

Fig. 66. A series of diagrams illustrating the method of introducing Barred Rock melanoblasts into White Leghorn wing skin, which is then retransplanted at hatching to the back of a White Leghorn host. (After Rawles.)

Nickerson showed that the presumptive tissue for the white bars actually contains melanoblasts which are capable of forming pigment under suitable conditions, and he suggests that their failure to do so in the feather, and thus the barring itself, is due to the gradual accumulation, during the phase of pigment formation when the black bar is produced, of an inhibitor which eventually stops the deposition of pigment, but in its turn leaks away or is destroyed during the pigmentless phase. The problems involved in the development of other two-colour patterns, such as the black-red patterns of Brown Leghorns and Rhode Island Reds, are more complex, since in these cases there seems to be an early determination of the melanoblasts into one or other of two different and persistent types of melanophore, one forming black and the other red pigment. This determination, which is concerned in the tract-specificity of the feather pattern produced, is carried out by the feather papilla, probably by the epidermal component.[1]

[1] For a full discussion, see Willier, 1942, 1948; and Willier & Rawles, 1944.

CHAPTER X

SOME GENETIC EFFECTS AND
THEIR PHENOCOPIES

FROM the earliest days of genetics, the domestic fowl has been a favourite material for investigation, both because of its convenience as an experimental animal and because of its economic importance. The genetic literature dealing with this animal is, therefore, very large, and no attempt will be made in this work to make a complete survey of it. A few genes are known, however, which produce extensive morphological abnormalities in fairly early developmental stages, and investigations of these give us some insight into the way in which genetic factors influence the developmental processes. Moreover, it has been found possible, in some cases, to produce morphological effects similar, to a greater or lesser extent, to those caused by certain genes, through the agency of various chemical or mechanical factors. There is again a very extensive literature dealing with external influences on development, such as the nutrition of the mother, conditions of temperature and humidity during incubation, effects of mechanical agitation, length of storage and similar factors. A review, including references to some 400 individual contributions, has been provided by Landauer (1941 c). Only those studies which are directly connected with the genetic effects will be mentioned here.

Rumplessness

Fowls lacking the tail feathers have been known for many years; and dissection reveals that in many of these the skeleton of the synsacral and tail region is affected, the whole of the latter being absent in the most extreme cases. The genetics of the condition was first investigated by Dunn (1925) and Dunn & Landauer (1925), who showed that, in the race with which they were dealing, it was due to a dominant gene. Much more recently, Landauer (1945 a) has described a second type of rumplessness in White Leghorns. In this the adults are in general rather similar to those of the previously known dominant strain, although they tended to exhibit greater distortion of the pelvis. The new condition, however, was genetically quite distinct from the old, and was inherited as a clean cut recessive.

The anatomy of the dominant conditions has been described in the papers by Dunn, and Dunn & Landauer cited above, and also by Landauer (1928). Its development was investigated by Zwilling (1942 b), following an earlier study by du Toit (1913). Zwilling found that in the normal chick embryo a rather large concentration of degenerating cells, characterised by the presence of chromophilic granules or globules in the cytoplasm, can be seen in the posterior part of the tail bud from the latter part of the second day (about 15 somites). In dominant rumpless chicks, the number of these cells is greatly increased, and they can be found further towards the anterior than in normals, so that by the end of the third day, when the process of degeneration is at its height, the whole of the indifferent cell mass of the tail bud and even the posterior end of the neural tube is affected. During the fourth day, when the most rapid growth of the tail should occur, the abnormality is clearly visible even macroscopically. Sections show that the tail somites fail to form, while both the notochord and nerve tube are shortened, the former usually ending in a thick blunt knob, while the latter often shows irregular cavities and bendings (Fig. 67, row 2).

In spite of the general similarity of the adult conditions, recessive rumplessness is produced in quite a different manner. Zwilling (1945 a) found that in this race tail formation proceeded normally until half-way through the fourth day. No excessive concentration of chromophilic cells is found in the early tail bud, but at the stage when the definitive nerve tube and somites should develop, they appear in abnormal form, the nerve tube being smaller than normal, and the somites frequently fused. The tail remains normal in length until the fifth day, when degeneration begins to occur. The extent of this is highly variable, but it often affects the neural system more than the notochord, which may stretch posteriorly considerably further than the nerve tube. In most of these cases, the vertebrae extend the full length of the notochord so that the skeleton of the adult may give no indication of the existence of abnormalities in the soft parts (Fig. 67, row 3).

Rumplessness has been produced in normal chicks by a variety of experimental methods. Danforth (1932–3) produced the condition by varying the temperature of incubation, and argued that it was caused by a general depression of growth during a critical phase. However, he did not subject his embryos to the temperature shock until 4–6 days of incubation, and Zwilling (1945a) argued from his studies of the development of the condition that by this time the critical phase should be over; he suggested that Danforth's results were not actually produced by the temperature

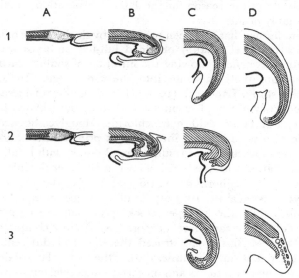

Fig. 67. Development of the tail: row 1 in normal embryos, row 2 in dominant rumplessness, row 3 in recessive rumplessness; column A at latter part of second day, column B at latter part of third day, column C at latter part of fourth day, column D on sixth day. The fine stippling indicates the 'indifferent cell mass', the coarse dots in rows 1 and 2 the region in which cellular degeneration occurs, the circles in row 3 show fused somites which underlie degenerating neural tube. (After Zwilling.)

treatment to which he attributed them. Landauer & Baumann (1943) found that mechanical shaking of eggs before incubation led to the production of rumpless embryos in a definite, though fairly small, number (c. 5 %). Nothing is known of the causal factors or developmental events in this case. More consistent results were obtained by Zwilling (1945b), who cut through the embryo in the region of the last somite, half-way between this and the tail bud or at the junction of the neural tube and the undifferentiated tail bud, in stages ranging from 11 to 32 somites.

A large number of embryos were produced which were exact phenocopies of the recessive rumpless condition, the incidence of them rising to over 60 % when the cut was made at the hindmost level. It is not unexpected that an operation made in the neighbourhood of the tail bud should cause abnormal morphogenesis of the kind found in this mutant, but it is perhaps more difficult to understand the reason for the later degeneration which must presumably occur.

Rumplessness has also been produced by the injection of various chemicals. Ancel (1945 b) obtained tail-less embryos with considerable frequency following the injection of sodium cacodylate or disodium methylarsenite into the 2-day egg. In a more extensive study, Landauer (1945 b) found significant increases in the frequency of rumpless embryos when cysteine hydrochloride, cystine, l-glutamic acid hydrochloride, thioglycollic acid and l-malic acid were injected into the yolk prior to incubation, but much the most consistent results were obtained with insulin. The response varied with the breed of fowls used; after the injection of 2 units, White Leghorns gave $41 \cdot 6 \pm 3 \cdot 8$ % of 17-day old rumpless embryos, as against $1 \cdot 5 \pm 0 \cdot 7$ % of sporadic rumplessness in controls, while the Creeper stock gave only $17 \cdot 1 \pm 2 \cdot 1$ % as against about the same level of sporadics. A more complete study (Landauer & Bliss, 1946) traced the effects of different doses, administered at different times during the first 2 days of development, on these two stocks and on certain lines selected from them. If the dose is expressed in logarithms, and the response in probits, a linear relation is found between the two over a considerable range. The Creeper response is always lower than that of Leghorns, but it could be shown that this is due, not to the Creeper gene itself, but to the general genetic background. There is, in fact, a difference in penetrance (to use the geneticist's term) of the insulin effect in the two stocks; and this could also be altered by selection, since a race of Leghorns was produced in which the frequency of rumplessness with a given dose was considerably increased. There were also differences in expressivity, in that the rumpless embryos induced in the Creeper stock were more often of the intermediate grade than those in the Leghorns. Finally, the injections also had a toxic effect, and the interbreed variations in respect of this were independent of those in respect of rumplessness, since in the selected Leghorns the 'low' stock showed a higher early mortality though a lower frequency of induced rumpless embryos.

The mechanism of action of insulin in this connection is still obscure. Landauer & Bliss showed that the maximum effect is produced by injections made at about 31 hours of incubation, and that the rumplessness-inducing effect falls off to zero at about 72 hours.[1] The development of the induced rumpless embryos has been described by Moseley (1947) and a full account of the adults has been given by Landauer (1947c). Moseley showed that there is considerable variation in the early developmental stages. In about two-thirds of the affected embryos, the caudal axial structures are bent sharply downwards and protrude into the gut, a condition known as ourentery. A certain number of similar cases were produced by Zwilling's (1945b) transections of the posterior part of the embryo; and some cases of chordentery (i.e. deviation of the chorda alone) are found to occur in control stocks and probably give rise to the sporadic rumplessness which seems to have a low frequency of occurrence in all breeds. Not all the insulin-induced rumplessness, however, can be attributed to ourentery, since nearly a third of the investigated embryos showed a condition which was very similar to that of recessive rumplessness; in fact, Moseley considered that most of the cases which survive to maturity are produced in this way. As well as the two main types, there are a few other abnormalities which could not be classified. Grünwald (1941, 1947b) has also described early developmental stages of sporadic malformations in which either the whole posterior end of the body protrudes downwards into the gut (ourentery) or in which only the chorda does so (chordentery). He draws attention to the similarity with some of the pictures seen in hereditary rumplessness, but points out the great variability of the phenomenon and concludes that it is impossible to reach any understanding of its epigenetic causation by a mere examination of the end-products. He argues (1947c) that in some cases such downward deviation of the axial organs may lead to the fusion of the posterior limbs and thus to the production of a 'sirenoid' monster.

There is thus a striking variability in the epigenetic events which lead to the production of the essentially similar rumpless

[1] At slightly later periods injections of insulin modify the expression of genetic polydactyly (p. 219) and at about 5 days they induced micromelia (p. 220).

adults found in the dominant and recessive mutant races, the mechanically produced phenocopies, and those due to insulin. The characteristics of the dominant mutant are well defined and do not seem to have been imitated by any other means. Clearly the dominant gene is involved in some specific way in the activity of the cells of the tail bud, which degenerate in considerable numbers if the dominant allelomorph is present. The effects of the recessive gene are presumably also to be attributed to the failure of some process which is usually mediated by the normal allele of its locus, but in this case the gene-controlled process would seem to be involved in some general complex of reactions which is altered in a similar way by a number of different factors, including transection of the axis, and excess of insulin. This is, of course, a common phenomenon, which underlies most cases in which phenocopies of gene effects can be induced by external agents. The peculiarity of the rumpless situation is that one of the most effective agents in producing phenocopies of the recessive mutant type, namely insulin, must also affect some epigenetic situation which is other than that for which this gene is essential, but which nevertheless responds by leading to a very similar end-result. This is the system through the agency of which insulin causes ourentery. One must assume that the metabolic processes affected by insulin are active in several different connections in the development of the tail bud, and that the insulin injections affect sometimes one of these (leading to recessive rumpless phenocopies), sometimes another (leading to ourentery).

The nature of these metabolic processes is still imperfectly understood. It is known (Zwilling, 1948 a, b) that at later stages injection of insulin produces hypoglycaemia. There is no direct proof that this is so for the early injections which induce rumplessness, but it certainly seems the simplest assumption. Moreover, Landauer & Lang (1946) showed that the teratogenic effect requires the presence of the complete insulin molecule, and is abolished and re-established when insulin is reversibly inactivated, as regards its effect on blood sugar, by treatment with iodine. Further, the rumplessness-inducing effect is reduced if substances (such as nicotinamide), known to antagonise the action of insulin on carbohydrate metabolism, are injected simultaneously with it (Landauer, 1948 b). A similar result is found in connection with the teratogenic effects of later injections which are known to cause hypoglycaemia; and Landauer (1949) has shown that such substances also protect the embryo against the induction of taillessness by sodium cacodylate described by Ancel, as well as against the later teratogenetic effects of sulphonamides, eserine, etc. (cf. p. 220). However, adrenal cortex hormone, which should increase hypoglycaemia, does *not* increase rumplessness with insulin, though it raises the incidence of the micromelia induced by injections of insulin at a later stage. In spite of this deviation from what might have been expected, it seems most probable that the rumplessness-inducing effect is brought about mainly by an influence on the carbohydrate metabolism.

Creeper

The so-called Creeper breed of fowls is characterised by extremely short and bent legs. The condition is due to a single dominant gene, which is lethal when homozygous (Dunn & Landauer, 1927; Landauer & Dunn, 1930). During the last 20 years a long series of investigations of this phenomenon have been made, chiefly by Landauer and his associates, and described in voluminous detail.

In the heterozygous Creeper, the main effects take place in the long bones of the legs, and are consequences of abnormalities in the development of the cartilage and its conversion into bone (Landauer, 1931). The cartilage cells are in some stages smaller, in others larger than in normal embryos; the periosteal ossification begins somewhat later, but progresses even further than usual, while endochondral ossification, although starting at the same time as in normal chicks, is arrested at an early stage. These changes are strikingly similar to those of the human congenital malformations known as chondrodystrophy. They lead to considerable reductions in the length of the various bones, which are also frequently bent.

The homozygous Creeper embryo is much more strongly affected. Landauer (1932) showed that they usually die on the third or fourth day of incubation, but a small percentage (some 2 %, depending on the stock) survive this critical period and remain alive till near the end of incubation.[1] These show extreme reduction in length of the limbs, great abnormalities of

---

[1] The proportion surviving is slightly increased if the first 24 hours of incubation are carried out at a reduced temperature (Landauer, 1944a).

chondrification and ossification, together with various defects in the eyes and other organs (Landauer, 1933). The shortening of the limbs resembles that in certain human malformations known as phocomelias and micromelias, and the survivors among the homozygous Creepers are usually referred to as 'phocomelic embryos' in contradistinction to the 'prothanic embryos' which suffer early death.

Landauer (1932) considered that the main abnormality exhibited by the prothanic embryos before their death was a general inhibition of growth, and he suggested that this was the prime effect of the Creeper gene; the specific effects on cartilage formation, eye-shape, etc., were regarded as reactions of particularly sensitive developmental processes to this primary general inhibition. Later investigation suggests that this view is actually almost the reverse of the truth, the fundamental effects of the gene being a relatively small number of specific modifications of development, of which the generalised inhibitions and growth retardations are secondary consequences: it seems probable, in fact, that eventually all the Creeper effects may be explicable as secondary consequences of a single primary effect, although it is not possible to carry the analysis so far at present.

The cause of death of the prothanic embryos was reinvestigated by Rudnick & Hamburger (1940). David (1936) had shown that they do not die in consequence of any lethal condition located in each and every cell, since he found that all tissues could survive long past the normal time of death when explanted into tissue culture. The lethality, therefore, must depend on some effect which is produced in the embryo as a whole. There is, of course, a difficulty in identifying such embryos in early stages, since they can only be produced by matings between two heterozygous animals, and therefore comprise only 25 % of a given batch of eggs. Rudnick & Hamburger counted the somites of embryos at certain times, and then continued incubation until the phenotype of the embryo became clear. They found very slight retardation of prothanic embryos in terms of somite number during the second day, but this had disappeared by the 48-hour stage. Cairns (quoted Hamburger, 1942a) made measurements of the brain vesicles of embryos stained with vital dyes, and, using continued incubation as a method of identification, showed that there is no significant difference in growth rate between prothanics and normals in the early somite stages. The same author (1941) made the significant observation that the prothanics entirely fail to develop a vitelline blood circulation. This condition is clearly recognisable at 54 hours of incubation (23–27 somites). Associated with this failure of the extra-embryonic circulation is the persistence of an anastomosis between the dorsal aorta and the anterior cardinal vein, by which the blood is returned direct to the heart. Hamburger (1942 a) regarded the continued existence of this anastomosis as the prime abnormality in the prothanics, but Cairns & Gayer (1943) interpret the situation in the opposite sense, considering that it results from the deficiency in the vitelline circulation. They found that in some prothanics a partial but defective vitelline circulation develops, and in these the anastomosis between the aorta and cardinal vein does not persist; it does not seem excluded, however, that it is a delay in the formation of the normal vessels in this region which leads to the inadequate establishment of the peripheral circulation. Whichever view is correct (and Hamburger's suggestion seems perhaps the more attractive), Cairns (1941) showed that if the vitelline circulation is interrupted by longitudinal cuts through the base of the vitelline arteries, the embryo suffers a series of modifications which exactly parallel those which precede the death of the prothanics. There is a general retardation of the body, with particularly strong inhibition of those organs which come to lie underneath, for instance, the left eye and otic vesicle, an enlargement of the heart and disturbance in its coiling, retardation of the amnion, and the collection of pools of blood within the embryonic vessels; many of these effects can also be seen, after the blood circulation has ceased, in embryos explanted *in vitro*.

The homozygous Creeper embryos which survive this critical period are presumably those in which just sufficient vitelline circulation develops. As shown by Landauer (1933) and Cairns & Gayer (1943) these embryos have extremely short and distorted limbs (phocomelia), a shortness of the lower jaw relative to the upper, rudimentary eyelids, microphthalmia, and coloboma, i.e. a failure to close shown by the choroid fissure, around the edges of which the retina is everted to lie on the outer surface where it replaces part of the pigmented layer. A variety of miscellaneous abnormalities which occur in the Creeper stock have been described by Landauer (1935 a).

Landauer (e.g. 1941 a) originally laid the greatest emphasis on the growth inhibition as the prime effect of the gene. He suggested that both in single dose, and still more in double dose, the Creeper gene causes a generalised retardation of growth at definite stages of development, to which the various organs react in a differential manner in accordance with their sensitivity at the periods in question. In pursuit of this line of thought, he studied in detail the growth of limbs and other organs both in heterozygous Creepers (1934a) and in homozygotes (1939a). Unfortunately the results are not presented in a way which makes it easy to grasp their significance, since Landauer discusses, not changes in the growth rates of the various elements, but changes in their actual lengths. He showed that in the heterozygote, the anterior limbs are more strongly affected than the posterior. Within the leg, the more distal bones suffer a greater retardation than the more proximal, although the toes are hardly affected. The distal bones are those which are laid down later in embryonic life (cf. Saunders, 1948, p. 170); they are also those with the highest inherent growth rate (Lerner & Gunns, 1937). According to Landauer, the importance of this inherent growth rate in determining the extent of the Creeper effect is shown by the fact that the longer a bone is normally, the greater is the absolute reduction in length; thus the femur, which is a short bone in the chick, is reduced by a smaller proportion than the much longer tibio-tarsus. In the homozygote, the effects follow a generally similar pattern, although much exaggerated; for instance, in this case the toes are strongly affected. The effect, however, touches not only their growth, but also their morphological pattern, since elements are frequently fused or otherwise distorted.

Landauer has described two other genes in the fowl which affect the growth of the long bones of the leg. One (1935 b, 1939 b) is a lethal mutation found in the Dark Cornish breed, the other ('short upper beak', 1941 b) a semi-lethal, which affects the leg bones and the length of the upper beak. In both these, the general pattern of growth retardation is very similar to that which had been described in heterozygous Creepers, which gives strong support to the hypothesis that all three genes affect, in a somewhat similar way, a basic pattern of growth which characterises chicks in general. Presumably other genes might be found which would also modify this pattern; but, impinging on it from some other angle, they would give rise to quite differently organised responses. No single genes of this kind have yet been described, but the genetic factors characterising the race, for instance, of Show Game fowl must, as a complex, have effects which are in some respects the opposite of those produced by Creeper, since in this race the legs are as much longer than normal as the Creeper ones are shorter.

However, it is open to question whether it is legitimate to speak of the Creeper effect as though it acted on some well-defined entity which we recognise under the name of growth. During embryonic life, increase in size of an organ such as a limb is intimately bound up, not only with increase in the numbers and volume of the cells composing it, but also with their histological differentiation. The Creeper gene certainly causes abnormalities in this last component, particularly in chondrogenesis and osteogenesis. During periods of development in which these or similar processes are actively under way, it is very difficult to define any entity which is simply growth unmixed with histogenesis; and if, as in this case, an alteration in final size is accompanied by abnormalities in the formation of the constituent tissues, it is impossible, from mere measurements, to class the effect as one on growth rate rather than as one on histogenesis.

Further insight has, therefore, been sought by experimental means. Fell & Landauer (1935) attempted to imitate the effects of the Creeper gene by imposing a growth restriction on normal limb buds, which were cultivated *in vitro* both in normal medium and in a medium which contained less than the usual proportion of embryo extract. In both these, a retardation of growth occurred, but this was more severe in the specially extract-poor medium. It was found that, among explants of any given stage (less than about 5 days), the proportion which showed successful ossification was always less in the growth-restricting medium than in the normal. In explanted mandibles the formation of membrane bone proceeded in all cases, so that the inhibition of ossification applies only to those types of osteogenesis which are preceded by a stage of cartilage hypertrophy. David (1936) also described the occurrence of chondrodystrophic limbs in chorio-allantoic grafts of normal limb buds. It remains obscure, however, as Hamburger (1942*a*) has remarked, whether these abnormalities of cartilage differentiation and bone formation are mere secondary consequences of the inhibition of growth in the isolation sites, or whether the sites are deficient in some substance which plays a specific part in these histogenetic processes.

Experiments have also been made to test the capacities of Creeper limbs when transplanted into normal surroundings. Hamburger (1939c, 1941) grafted both heterozygous and homozygous Creeper limbs into the coelom of normal hosts. Previous work by Hamburger & Waugh (1940) had shown that, under such conditions, normal limb buds develop into comparatively normal limbs, which are, however, about 20 % shorter than normal. The growth inhibition which must have affected these control limbs is ascribed to general, non-specific deficiencies such as inadequate vascularisation and nerve supply. The transplanted heterozygous Creeper limbs were found to be reduced below the usual Creeper size by an approximately similar amount; but there was considerable variation and some were much smaller. Hamburger emphasised the fact that there was not, as might have been expected, a parallel series of gradations by which the transplanted normal limbs became more like Creepers, and the reduced heterozygous Creepers more like the phocomelic homozygotes. In fact, the growth inhibition exerted by the coelom transplantation-site seemed to produce little or no effect on histological differentiation. This reduces the plausibility of the view that the characteristic Creeper effects are secondary consequences of a primary growth modification. With more certainty, the transplantation results demonstrate that the deficiency, of whatever kind it may be, which is caused in the tissues of the limb bud by the Creeper gene, is an 'autonomous' one which cannot be made good by the diffusion of a missing substance from the neighbouring normal tissue. Hamburger's earliest grafts were made from donors with some nineteen pairs of somites, but Rudnick (1945a) obtained essentially similar results with grafts taken from embryos as young as the 6-somite stage, which demonstrates that the Creeper defect cannot be overcome even at a stage before the appearance of the limb bud.

Studitsky (1944, 1946) has claimed that when the tibiae of Creeper heterozygotes $7\frac{1}{2}$–8 days old are grafted on to the chorio-allantoic membrane of normal hosts they develop into perfectly normal limbs which show no chondrodystrophy; he also finds that this condition can be induced in normal embryos of similar age by the presence of pituitary gland grafted on the chorio-allantois. Landauer (1946) has pointed out the unconvincingness of the evidence offered by Studinsky in support of his first point and there seems no doubt that his evidence cannot be accepted in face of the much more thorough and technically perfect experiments of Hamburger and Rudnick. It is also certain that the abnormalities of homozygous Creepers are visible long before the pituitary starts to function, which makes it appear very improbable that hormonal agents have anything to do with the deformities found in heterozygotes.

The discussion has so far dealt mainly with the effects produced by the Creeper gene on the limbs, but it has been mentioned that in the homozygotes several other organs are also abnormal, particularly the eyes, which are not only smaller than normal but exhibit the condition of coloboma. The formation of the scleral cartilage is also deficient and the choroid coat is incomplete. Gayer (1942) transplanted optic vesicles from normal and from phocomelic embryos of 9–24 somites into the coelom of the flank of normal hosts. He found that both types of isolate developed typical coloboma, with much the same range of variation in the severity of the condition. Gayer & Hamburger (1943) then grafted homozygous Creeper eyes into the site from which the eye had been removed in normal hosts, and found that under these conditions the transplanted eyes often developed as normals, showing no sign either of the microphthalmia or of the coloboma which might have been expected. The few normal eyes which survived long enough in homozygous Creeper hosts were colobomatous. They suggested from these results that the Creeper gene is ineffective within the tissues of the eye itself, the typical defects of the eyes in homozygotes being impressed on the optic cup by influences exerted by the surrounding mesoderm. The mesoderm of the head of normal embryos allows the optic vesicle, whether Creeper or not, to develop fully, while the homozygous Creeper head mesoderm, they suggested, impresses the defective character on the intrinsically adequate eye-cup developing within it, just as does the flank mesoderm.

Landauer (1944 b) pointed out that the influence of the surroundings might not be by means of an inhibitory action which can induce defects, but through a failure in the supply of substance which is necessary for eye development, and which is produced in normal head mesoderm but not in either flank mesoderm or homozygous Creeper head mesoderm. If that were the case, it is possible to suppose that the homozygous Creeper eye-cups are in fact defective, but their deficiency is not too great to be overcome by material supplied by the head mesoderm. It is unexpected, on this hypothesis, however, that normal and homozygous Creeper eyes should exhibit such similar capacities when transplanted to the flank; and, in general, while Landauer's suggestion does not seem definitely ruled out by any facts at present available, it seems more complicated than the original suggestion of Hamburger & Gayer. These authors also produced evidence for the normality of the homozygous Creeper eye, in a rather different connection it is true, by demonstrating that although when in place in its own head it does not become surrounded by a normal scleral cartilage, it is quite able to induce the formation of this covering when transplanted into normal hosts. Thus here there is no doubt that the defect characteristic of the Creeper condition is due to an inadequacy in the head mesoderm, which should react to the inducing stimulus, and not to an inadequacy in the eye, which should exert it. This induction occurs, however, even when the homozygous eye is transplanted into the flank and develops coloboma, from which one must conclude that the deficiencies involved in the two abnormalities are not very intimately related.

The actual cause of the death of the homozygous embryos which pass the critical phase in the third and fourth days is not very easy to decide. The phocomelia and the other abnormalities such as coloboma would obviously unfit them for life outside the shell, but in fact they never hatch, and die some time towards the end of the normal incubation period. Landauer (1939b) showed that they suffer from a grave anaemia, probably as a consequence of the absence of bone marrow in the extremities; and there is considerable compensatory hypertrophy of the heart and spleen. But it seems unlikely that it is the anaemia which brings about their final death.

The various effects of the Creeper gene have been summarised by Hamburger (1942a) in a diagram which is reproduced in Fig. 68.

A certain number of physiological studies have been made on the Creeper condition. Creeper heterozygotes react more strongly to lack of vitamin D than their normal sibs (Landauer, 1934b)

and their rickets differs in its histological details from that of the normals. The Creepers are not affected by castration, which is also without effect on the bone growth of normals (Landauer, 1937). Their content of glutathione and ascorbic acid is normal, while that of homozygous Cornish lethals is reduced; homozygous Creepers were not assayed (Gregory, Asmundsen, Goss & Landauer, 1939). Franke and co-workers (1936) have shown that, if fowls are fed a diet containing traces of selenium, they

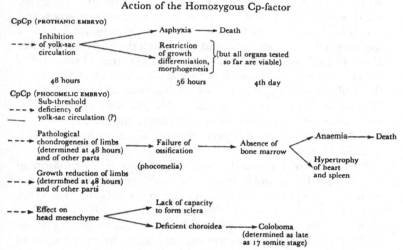

Fig. 68. Diagram of the action of the homozygous Creeper factor on prothanic and phocomelic embryos. (After Hamburger.)

produce eggs in which the embryos show deformities of the head and extremities. The progeny of heterozygous Creeper fowls show the effect more strongly than those of normals, presumably because the selenium intoxication exaggerates the effects of the heterozygous gene (Landauer, 1940).

Polydactyly

The legs of most breeds of fowls contain a metatarsus which consists of three bones fused together, to which is located a series of four digits, the first being the hallux, which is attached somewhat more proximally than the other three. The occurrence of a five-toed, or polydactylous condition has been known for very many years, and is typical in certain breeds such as Dorkings, Silkies, Houdans, etc. Very soon after the rediscovery of Mendelism, Bateson (1902) showed that the polydactylous condition is determined by a single main gene which is dominant over that for the four-toed leg. The gene concerned is now known as P_0. It produces some effects on the wings, which have been described by Baumann & Landauer (1944). More recently, at least three further polydactyly-producing genes have been described. These are: (1) Another dominant form termed 'duplicate' (Warren, 1941, 1944), which was found in a White Leghorn stock and which has been shown to be an allele of the original P_0; (2) a recessive lethal condition 'talpid', in which the embryos die after about 8–10 days' incubation, showing duplication of toes and of digits of the wing, often with considerable reduction in the size of the limbs (Cole, 1942); (3) another lethal condition 'diplopod' (Taylor & Gunns, 1947), which is also inherited as a recessive and causes death shortly before hatching, the chicks having a considerable number of supernumerary digits on both the fore and hind limbs.

Pure-breds of the breeds containing P_0 are fairly uniform in appearance, but on outcrossing to normals, considerable variation is found in the polydactylous segregates in the progeny. The other mutant types are always highly variable in expression, and it is clear that the condition is affected by numerous modifying genes. Perhaps the least severe expression consists in the enlargement of the hallux, which contains three phalanges instead of the usual two. Warren (1944) suggests that a series of increasing effects involves, first, the addition of a longer toe outside the hallux, then the splitting of this digit, and finally the loss of the normal hallux. On this basis it might be suggested that the polydactyly effect consists essentially in the addition of extra digits to the pre-axial side of the normal foot. However, Warren's series does not cover some of the more extreme grades, such as those produced by diplopod, or those described by Harman & Nelson (1941). The development of such types was studied at a very early date by Kaufmann-Wolf (1908) and Braus (1908), who both called attention to the occurrence of a tendency for the polydactylous foot to be bilaterally symmetrical, the pre-axial part being a mirror image of the post-axial. This type of symmetry relation is well known in limbs of, for instance, Amphibia, which have become reduplicated following surgical operations. These authors, therefore, suggested that polydactyly represents a partial reduplication of the foot, and this suggestion is accepted by most recent authors, such as Landauer (1948*a*), Gabriel (1946*a*) and Grünwald (1947*a*).

There can be no doubt that even in a foot subject to this tendency to reduplication, the post-axial portion retains a dominant position. One must suppose that within the footplate of a developing limb bud there exists a system of interactions (perhaps dependent on diffusion phenomena) which cause the mesenchyme to condense into four rods which are the precursors of the digits. Such a system is a typical example of an 'individuation field' (p. 124). The effect of the polydactyly genes is to

Fig. 69. Some types of polydactyly in adult fowls. All are left feet except that on the extreme right of the row. (After Landauer.)

alter one or more of the reactants in such a way that they no longer reach equilibrium with one another so as to result in the formation of the normal pattern. It is the pre-axial part of the pattern which is primarily affected. In the extreme grades, this region tends to form a complete new system of condensations with mirror symmetry to the normal system in the post-axial region. The milder grades show only an enlargement (hyperphalangy) of the hallux, or the addition to the pattern of a supernumerary element. If one adopts the hypothesis of reduplication consistently throughout the whole series of grades, one would have to suppose that this extra digit beyond the hallux is always to be considered as representing the fourth toe. This seems rather an artificial assumption, and, on the basis of the physiological considerations just mentioned, there is no theoretical necessity for it. We could suppose simply that the polydactyly genes upset the normal processes of pattern formation as they affect the pre-axial part of the footplate (possibly, for instance, by slowing down the spread of inhibitory influences from the post-axial region, or by increasing the mass of pre-axial tissue beyond what the normal influences can control), but that it is not until this disturbance reaches a certain minimum degree that the pre-axial portion is able to produce within itself a mirror-imaged replica of the normal pattern of condensations. There need be, therefore, no inconsistency between the two suggestions that polydactyly is due to the addition of elements to the pre-axial side or that it is due to partial reduplication.

The condition of the apical region of the limb bud, which Saunders (1948) has shown to play such an important role in the development of the wing (p. 170), does not appear to have been investigated in polydactylous embryos.

Phenocopies of polydactyly can be produced by the surgical division of very early rudiments (Wolff & Kahn, 1947a, b; cf. p. 168). The expression of the condition in genetically polydactylous stocks can also be considerably modified by experimental procedures. Both Sturkie (1943) and Warren (1944) have shown that low temperatures of incubation partially suppress the condition and that the effect is greatest during the third day, disappearing after the fifth. No very certain conclusions as to the mechanism of the gene effect can be drawn from this, but it suggests that a lowering of the growth rate in the developing limb bud is antagonistic to the formation of a polydactylous limb. Gabriel (1946a) came to what at first sight seems an opposite conclusion from the results of experiments in which parts of the limb bud were treated with small plates of agar soaked in colchicine. Treatment of normal buds with this alkaloid may completely eliminate the limb, and there is no doubt that the prime effect of the substance is inhibitory, operating through the well-known arrest of mitosis which it causes. Gabriel found that, if the whole limb bud or the pre-axial portion of a polydactylous bud is treated, the expression of the gene is reduced, as might be expected. However, treatment of the post-axial region alone led in some cases to a considerable exaggeration of the polydactylous effect. He draws the conclusion that it is the relation between the pre- and post-axial regions which is the important factor in the determination of polydactyly, and thus arrives at a view of the physiology of the condition very similar to that described above, except that he assumes that any release of the pre-axial region from the general individuation field will necessarily lead to a tendency for this portion to produce a complete new field and thus a reduplication.

Modification of the expression of polydactylism can also be obtained by the injection of insulin (Landauer, 1948 a). The most sensitive period is at about the third and fourth day, that is to say somewhat later than the period during which rumplessness is produced, and before the period at which the injections cause the 'short upper beak'-micromelia syndrome (p. 220). The response is highly variable, but is always in the nature of a suppression of the abnormality. Landauer found that the extent of the insulin effect varied greatly in different stocks. Polydactyls belonging to the pure breeds, which have been selected for polydactyly for long periods, showed much less effect of insulin treatment than did polydactyls which had segregated out from crosses with unrelated stocks. The former are also less variable than the latter when both are untreated. As Landauer points out, this shows that, during the formation and maintenance of the polydactylous breeds, many modifying factors, which are present in normal breeds and are capable of altering the grade of expression of polydactyly, have been eliminated.

A remarkable feature of the polydactylous condition is the relative frequency with which only one limb is affected, or affected more strongly than the other. Such 'heterodactylism' has been investigated by a number of authors, whose work is reviewed in Landauer (1948 a). The condition is most often found in hetero-zygotes, and in unselected stocks the most strongly abnormal limb is much more frequently the left (a situation which is usually, though somewhat unhappily, spoken of as 'sinistral heterodactyl-ism'). The frequency of the sinistral type can be raised fairly rapidly by selection, but that of the dextral type does not respond so readily, although there is evidence that hereditary factors play a role in determining its occurrence. Some of the older authors (e.g. Bond, 1926, 1932) suggested that the unilateral expression of the polydactyly gene was caused by a differential distribution of the gene during early development; as 'a new factor or gene', it was supposed to 'fuse with difficulty with the older established com-plex', and thus frequently be lost during early cell divisions. Such a suggestion is very far from our present ideas of gene behaviour,[1]

[1] Genetic mosaics are, of course, known in many organisms and occur in the fowl (reviewed in Crew & Munro, 1938, 1939; Greenwood & Blythe, 1951).

and Landauer makes the much more attractive suggestion that heterodactylism is a result of the well-known asymmetry of the chick embryo, and is therefore related in some way to the rotation of the body on to its left side. Not only are the factors bringing about this rotation themselves very inadequately understood, but likewise it is quite unknown what are the particular consequences of it which influence the development of polydactyls.

Micromelia and 'short upper beak'

A number of chemical agents are known which produce abnormalities in cartilage formation, leading to the production of embryos with reduced limbs (micromelia) and often with defects in the beak as well. Effects of this kind have been reported both in riboflavin deficiency (Romanoff & Bauernfeind, 1942) and biotin deficiency (Cravens, McGibbon & Sebesta, 1944). They were also obtained by exposing 48-hour embryos to the action of certain sulphonamides (Ancel & Lallemand, 1942; Ancel, 1945 a), while the effects on the beak are found after treatment with physostigmine (Ancel & Lallemand, 1844). Rather similar abnormalities are caused when the laying hens are fed on a diet deficient in manganese (Byerly, Titus, Ellis & Landauer, 1935; Lyons & Insko, 1937); but these were shown to differ histologically from the conditions characteristic of the usual chondrodystrophy (Landauer, 1936).

The most thorough study of such effects, however, has been made by Landauer and his associates, using insulin as the agent. It has been pointed out that when insulin is injected during the first 2 days of incubation, it induces a condition of rumplessness; if injected during the third and fourth days its effects are related to polydactyly, while injections on the fifth day produce micromelia and malformations of the beak. The shortening of the limbs is not unlike that found in Creeper, but the whole syndrome is more closely parallel to that caused by a gene known as 'short upper beak', which Landauer (1941 b) extracted from a Houdan crossbred flock. This gene is a recessive and originally produced extreme inhibition of the growth of the maxilla and of the long bones of the limbs, usually resulting in death before hatching. Continued selection against the expression of the trait, however, led to the formation of a stock in which the effects of the gene are much less severe and the homozygotes usually viable. The mechanism by which the gene, or the injected insulin, produces defects simultaneously in the upper beak and in the cartilages of the long bones is not understood; but the fact that the genetic effect is exactly paralleled by the insulin abnormality strongly suggests that there is some physiological process common to the two phases of development, and that the apparently different (pleiotropic) effects of the gene are really secondary consequences of some single initial action.

Embryos of a 'short upper beak' stock which had been selected for some generations for low-grade expression of the character, gave a much lower frequency than do normal stocks of shortened extremities when insulin was injected after 3 days of incubation, and were also less liable to rumplessness following early injections (Landauer, 1947a). Similarly, in Creeper stocks, in which some selection for suppression of the character has usually occurred, the reactions to insulin are comparatively mild. It has already been pointed out (p. 219) that

Fig. 70. Insulin-induced shortening of the upper beak and micromelia in a White Leghorn embryo. (After Landauer.)

in the selected, pure-bred polydactyls the pattern of the foot is less readily modifiable than it is in polydactyls derived from outcrosses. These are good examples of the phenomenon which has been called by Waddington (1940a) the 'canalisation of development', that is to say, the production under the influence of natural or artificial selection, of a system of genes which interact in such a way that the developmental system which they set in motion is relatively invariant under the influence of slight abnormalities, whether environmental or genetic. The normal development of animals subject to natural selection is usually highly canalised; an example from the present material is provided by the fact that doses of insulin which modify the pattern of a polydactylous foot make no effect on a normal foot. But the canalisation (or 'buffering', to use an alternative expression) of the polydactylous pattern can be increased by selection, and the relatively buffered pattern is then less sensitive to modifying influences than the pattern of an animal in which the genetic background has not been adjusted to canalise the effect of the major gene. As Landauer (1947a) points out, the evidence suggests that the modifying genes which effect the canalisation of one pattern (for instance, short upper beak) may have a similar effect on some anatomically unrelated pattern, since the selected short upper beak stock is also less liable to insulin-induced rumplessness.

The mechanism of the insulin action in inducing the various abnormalities which it can cause at different stages of incubation has been the subject of some investigation, although one cannot claim that it is yet fully understood. Landauer (1948c) has recently given a general discussion. Zwilling (1948a, b) has shown that insulin, injected on the fifth day, causes a considerable fall in blood sugar values, which is surprisingly long lasting, being apparent until the fourteenth day. Adrenal cortex extracts (cortin) in themselves cause a mild hypoglycaemia, and strongly increase the effect of insulin when used in combination with it. Landauer (1947b) had previously found that injections of cortin combined with those of insulin caused the induction of more frequent and more severe cases of micromelia than did a similar dose of insulin alone. Epinephrine was without effect when injected along with insulin. Following the theory that the teratogenic action of insulin is a direct result of interference with carbohydrate metabolism, Landauer (1948b) injected simultaneously with it some substances thought likely to play an important role in embryonic metabolism. He was successful in demonstrating that nicotinamide exerts a considerable protective action against the micromelia-inducing influence of insulin, while α-ketoglutaric acid acts in the same way but less strongly. Nicotinamide exerts a similar protective effect against the micromelia-inducing action of sulphonamides, which may therefore act, like insulin, through the carbohydrate metabolism (Landauer, 1949). Further, Ackerman & Taylor (1948) found that micromelia is induced in eggs into which they had injected 3-acetyl-pyridine, a substance known to be antagonistic to the activity of nicotinamide. Zwilling and DeBell (1950), however, state that these abnormalities are unlike those of typical insulin micromelia. Sulphanilamide, on the other hand, produces results very similar to those of insulin, although it differs in being most effective when injected as early as 48 hours. They advise some caution in accepting the theory that the induction of micromelia is due to interference with the carbohydrate metabolism but, in the present state of our knowledge, that appears still to be the most probable hypothesis. If it is accepted, this is an instance in which an aspect of metabolism, which at first sight appears purely concerned with the maintenance of the embryo and the supply of energy, actually turns out to be closely involved in the morphogenetic events.

It is not certain, though probable, that the rumplessness-inducing action of insulin operates through the same mechanism; the question is discussed on p. 206.

CHAPTER XI

THE DEVELOPMENT OF THE EPIGENETIC
ORGANISATION OF THE CHICK

THE concept of organisation is a troublesome one to discuss adequately. That animals are not mere haphazard assemblages of parts is one of the most obvious and striking facts of even the crudest biology. But the word 'organisation', which implies some kind of contrast to such conditions of random aggregation, is often used either very loosely, without specification of what kind of order it is intended to denote, or too operationally, as though organisation could be invoked as an overriding power which could control the activities of the particular parts of the developing embryo. Both these errors can and must be avoided in any discussion which merits being considered scientifically worth while.

Organisation is clearly a matter of the relation between the whole of a thing and its parts. However, parts may be derived by any of an unlimited number of ways of division, and there are many kinds of relation between them and the whole. The organisation of an entity therefore has a precise sense only in relation to some specific sphere of relevance which defines both the type of derivation of the parts and the nature of the part-whole relationships. If, as is usually the case, the 'parts' in question are actually conceptual components, such a method of derivation can be considered as a path of abstraction. One must therefore always specify, in considering organisation, the path of abstraction, or, what amounts to the same thing, the sphere of relevance, in connection with which it has been divided into parts. Only in the most general philosophical discussion is it appropriate to consider organisation in general, when all possible modes of deriving the parts would be of equal interest. In normal scientific discussions one is concerned with organisation of some particular kind; and this may be defined as the relation, in connection with some particular sphere of interest, between a whole and the parts into which it may be divided (Waddington, 1949).

In embryology, one is most usually concerned with two such spheres of interest, which might be crudely referred to as the fields of descriptive and experimental embryology respectively. Under descriptive embryology one would consider the morphological or anatomical organisation of the germ—a matter not far removed from the purely geometrical, and which is the subject of descriptive developmental texts. In this book, however, we have been dealing rather with the second type of analysis, in which the whole embryo is divided, not into mere shaped masses of tissues, but into sub-units which are causally effective in influencing the sequence of events which constitutes the epigenesis of the developing animal. The sub-units derived in this way may sometimes be spatially defined entities, as for instance the lens-inducer is the eye-cup; but in other cases the epigenetic parts may be entities such as gradients, diffusion fields or the like, which need not be co-terminous with any particular anatomical organ, and in any case have a content which cannot be summarised in purely geometrical or anatomical terms. The relation of these sub-units to the whole of which they are a part constitutes the epigenetic organisation of the embryo. It is the aim of this section to summarise what has been said or implied about this in the earlier chapters of the book.

The first stage of the chick embryo at which we have any knowledge of the epigenetic situation is that at about the time of endoderm formation, just after laying. The most definite and important facts concerning this period are those derived from Lutz's duplications produced by sectioning the germ and Waddington's duplications and alterations in axial direction produced by mutual rotations of epiblast and hypoblast. Lutz's experiments clearly show that the blastoderm at this time is still organised as a single entity, at least with respect to the tangential plane. That is to say, the epigenetic fate of any plane part of the blastoderm (obtained by sectioning vertically to the tangential plane of the ovum, in which plane the blastoderm lies) is a function of its relation to the whole. Not only is this fate altered if the part is separated from the whole, but the part then usually itself develops into a whole embryonic area. There is then, within this plane, no separable sub-unit smaller than the whole: the blastoderm is a unity, and any part of it can become a similar unity. That does not mean, of course, that the unity is epigenetically featureless. There are clear indications of localised differences within it. There is, for instance, a polarity which determines which is to be the anterior and which the posterior end. Moreover, it appears that this is more strongly fixed in the posterior region, since, on sectioning, it is always this part which retains its original polarity while that of the anterior region may change. There may be a gradient in the firmness with which the polarity is determined, and it is also possible that the polarity is itself the expression of a gradient of some unknown character; but the situation is still obscure and will remain so until means of controlling the polarity of the isolated parts of the blastoderm have been discovered.

In the plane perpendicular to the blastoderm (the plane radial to the yolk on which the blastoderm lies) the anatomical unity of the blastoderm is breaking down, through the separation of the endoderm, at and just after the stages used by Lutz. Waddington's results show that the epigenetic unity also breaks down in this plane. By the time the endoderm becomes a separate layer it possesses its own polarity and is able to impose this on to the epiblast. When the experiments were made, the epiblast had also acquired a separate polarity; and, although the evidence that it can impress this on the hypoblast is not quite so clear, it is certain that at a later stage it can induce the formation of a fore-gut. The evidence that the hypoblast can, after its rotation, influence or even induce the form-building movements which lead to the appearance of the primitive streak suggests, however, that it may be this layer which at first plays the leading role.

The organisation of the avian blastoderm at this stage is comparable in many respects to that which is known in other vertebrates. Divisions vertical to the almost flat plane of the blastoderm correspond to divisions perpendicular to the spherical surface of the eggs in amphibians. As is well known since the work of Spemann, such divisions at the blastula stage may give rise to duplications, just as Lutz's did in birds. A similar form of regulation has been demonstrated in early stages of fish embryos and the studies of Glücksohn-Schönheimer (1949) on the mutant *kinky* makes it very probable that the same situation is true of mammals of a similar age.

There are some ways, however, in which the results with birds differ from those in other groups. Not enough is known of the epigenesis of mammals to offer a profitable field for discussion. We know slightly more about fish embryos. The extraordinary influence of the yolk on the developmental capacities of isolated blastoderms (Oppenheimer, 1934, 1936*b*; Tung, Chang & Tung, 1945) occur at an earlier stage than any which have been studied in birds, and much as we should like to know whether essential substances pass from the yolk into the early blastomeres in avian development as they do in that of the teleost, we are still completely ignorant on the point. At least by the time of laying any such transfer must be complete. More comparable, perhaps, with Lutz's experiments are those of Luther (1936, 1937), in which small pieces of the germ ring of the trout were isolated and cultivated in the embryonic shield or on the extra-embryonic membranes of host embryos. It was found that tissues characteristic of all parts of the embryo could be formed from any part of the blastoderm, indicating a unitary organisation similar to that of the chick. By contrast, Oppenheimer (1936*a*), working on *Fundulus*, found that the embryonic structures were only formed from the region near (within 45° on either side of) the presumptive embryonic axis. This may represent a real difference between these two species of teleost, but alternatively it may be a consequence of the technical limitation of the methods available. It has been seen that Dalton (1935) claimed, as a result of chorio-allantoic grafting, that in the chick also only the posterior quadrant of the blastoderm could produce embryonic structures. This result is certainly due rather to technical deficiencies in the test method than to real limitations in the developmental capacities of the material tested, although, as pointed out above, it is probable that the posterior, presumptively embryonic, region has some advantage in the degree with which the epigenetic polarity is fixed in it.

More detailed comparisons can be made between the avian and the amphibian embryos. In the latter, regulation following division along a radial plane is of course well known, both when the division is made soon after fertilisation or at the late blastula or very early gastrula stage, which corresponds more closely to the phase of avian development which is at present under discussion. There are, however, several differences to be noted between the behaviour of the two forms, as far as they are known at present. In the first place, it has been shown that the early amphibian gastrula can regulate to form a single complete embryo not only when material is removed from it, but also when considerable quantities of tissue are added to it (Waddington,

1938 *b*). There is no evidence that the avian blastoderm has the same ability, but equally there is no evidence against it, since the experiment has not yet been performed.

Of more significance are those differences for which there is positive evidence. There are two important ones, which are in some ways related. In the amphibians, the classical experiments of Ruud & Spemann (1922) indicated that only the presumptive dorsal half of the blastula is capable of forming an embryonic axis after isolation, and that axis is always more or less in the original direction: the ventral half-blastula formed only a relatively undifferentiated 'Bauchstuck'. Newer data of Dalcq & Huang (1948) have somewhat complicated the picture, since they show that the position of the blastopore may be considerably influenced by the reaction of the germ to the operation, some invagination occurring from the cut edge. But even these new observations do not suggest anything comparable to the complete development of a supernumerary embryo, often with reversed polarity, in isolated anterior fragments which was often seen in Lutz's experiments on avian blastoderms.

The other main difference is related to the former in so far as it also indicates a greater lability in the organisation of the avian blastula than exists in the amphibian. In the latter group there is no reason to believe, and much to deny, that the blastula endoderm can in any circumstances induce the formation of a new site of mesoderm invagination, as the rotated hypoblast did in Waddington's experiments. For a comparable influence in the Amphibia, we might perhaps look to the role played by the heavy yolky endoplasm in the unsegmented egg, as shown in the old 'Umdrehung' experiments of Schultz and their more recent repetition by Pasteels (1938–41). We have here an internal mass which plays much the same part in fixing the position of invagination as is performed in the bird by the hypoblast. The latter is also an 'internal' mass, which does not touch the external surface of the germ. But not only are its epigenetic activities exerted at a much later stage than the similar processes in the Amphibia; the disposition of the internal mass is also different. The deuteroplasm of the Amphibia forms a single mass within the spherical egg, whereas the hypoblast, at its first appearance, forms a circular ring wider at the presumptive posterior side, underneath the flat blastoderm. To be comparable to this, the amphibian deuteroplasm would have to take the form of a ring extending right round some parallel of latitude of the egg: if it did so, one might expect to find that an isolated ventral half of the egg would form a new blastopore in the ventral 'Randzone' and thus give rise to a secondary embryo of reversed polarity comparable to those described by Lutz. It is apparent, however, that the amphibian deuteroplasm does not extend right round the egg, or at least does not do so in sufficient quantity to induce a ventral blastopore in the isolated ventral half. Thus the two main differences in the behaviour of amphibian and avian blastulae mentioned above seem to find their explanation, first, in the fact that the epigenetic determination of the site of invagination, due in both cases to the action of an internal mass, takes place at a later stage in the bird; and secondly in the circumstance that this internal mass is more asymmetrically located in the amphibian.

The reasons for the temporal difference between the epigenetic systems in the meroblastic and telolecithal eggs are, of course, still uncertain, but it may be that the original structure of the oocyte is, in the latter more or less swamped by the enormous quantities of yolk which are added to it, so that it persists only in a weakened form which requires a considerable lapse of time before it becomes expressed in the development of localised embryonic structures.

The foregoing account of the first stages of gastrulation is sufficient to indicate certain differences which must inevitably exist between the organisation centre of the amphibians and the primitive streak of the chick. Thus the latter must lack the endodermal component found in the former; but this perhaps makes little difference in their epigenetic performances, since even in the amphibians it is the mesoderm rather than the endoderm which plays the greater part in inducing the remainder of the axis. Further, in the amphibians the direction of the main gastrulation movements is the same as that of the elongation of the embryonic axis while in birds the stream of invaginating mesoderm moves more or less at right angles to the axis of elongation. This has consequences on the details of the structures induced by the two organisation centres, but it is not a matter which raises any points of theoretical principle. It is perhaps of more importance that in the chick the first invaginated mesoderm is presumptive lateral plate, which in the amphibians shows less inductive power than does the dorsally located axial mesoderm.

It is difficult at present to say how far similar conditions apply in the chick. We have little information on the relative inductive efficiencies of the early streak, when it is made up mainly by lateral plate mesoderm, and of similar regions of the late streak when they are occupied by axial mesoderm; the experiments of Woodside (1937) are not very convincing, and do not make any allowance for the effects of different lengths of exposure of ectoderm to the inducing stimulus (p. 108). On the other hand, it is certain (Waddington, 1932 *et seq.*) that in the fully grown streak stage, the anterior part of the streak, which contains mainly axial mesoderm, is a more potent inductor than the posterior part. But this information should be completed by data on the inductive powers of presumptive mesoderm located laterally to the streak. The results of Abercrombie & Bellairs (1951) demonstrate that these lateral regions are capable of exerting an action on the polarity of the developing axis, and it seems most probable that they are capable of induction. This would not be surprising in view of their presumptive fate; but no direct experiments have yet been made to test this.

In the general type of inductive action which they perform, the amphibian organisation centre and the chick streak seem closely similar. Both can induce axial tissues, such as neural plate, chorda and somites, and in both cases the induced material may sometimes remain quite separate from the inductor, and at other times be united with it into a single axis, partial or complete, a distinction which Mangold (1932) emphasised in his terms 'complementary' and 'assimilative' inductions. Both organisers, again, can perform dynamic inductions, causing the reacting tissues to carry out certain morphogenetic movements which they would otherwise not have done. Finally, in both there is evidence of some degree of regional differentiation which, however, is still rather labile at the time when invagination is occurring.

It remains to be discussed whether the evidence accumulated from isolation experiments, particularly those of the American 'chorio-allantoic' school, indicates any important differences between the epigenetic situations in the chick and the amphibian. Those experiments have been discussed in some detail earlier (p. 77), and it was concluded in the first place that they are reconcilable with the view that neural differentiation is exclusively a result of induction by the mesoderm, and secondly that they gave evidence of a gradually progressing localisation within the blastoderm, at first of the region in which mesoderm can be formed, and later of the areas from which different types of mesoderm may arise. Such results are completely parallel to those which have been revealed in the amphibians, although in that group there is, as we have seen, a longer gap between the localisation of the presumptive mesoderm and that of the regions within it, while in the chick these two phases pass more continuously into one another.

The American authors have tended, particularly in the earlier years, to express their results in terms of a set of concepts which apparently differ rather sharply from those of the Spemann school. Lillie (1929) introduced the idea of 'embryonic segregation', which was developed in this connection by Hoadley (1926 a, b; 1927), and by Willier and his pupils. As it was at first used, this idea contained two quite distinct parts: the origin of new properties in a given region, and the progressive restriction of an originally highly diverse set of capacities. The inadequacies of this muddled concept has been discussed elsewhere (Waddington, 1932, 1935, 1940) and it is only necessary here to summarise shortly the way in which the classical Spemannian outlook on epigenetics would express the phenomena in question. In the first place, some new capacities undoubtedly arise. These are primarily competences, that is, states of instability during which several courses of future development are open to the tissue in question. The resolution of the instability constitutes determination, and this corresponds to a 'restriction of potency'. This determination may take place as a response to a stimulus coming from outside the mass of tissue, in which case we have to do with an inductive reaction. On the other hand, the instability may be gradually resolved by some autonomous process internal to the mass; and in that case it is found that, if the mass is large enough, the instability is differently resolved in different parts of it. This is the process which occurs in the original localisation of the organisation centre within the germ, and in the gradual appearance of regional specificities within that centre; it may also be operative in the acquisition of regional specificity within the neural system, though here inductive influences cannot be excluded. It can properly be referred to as 'segregation', if one so desires, although that term may seem to imply the spatial sorting out of actual substances which were originally mixed, and thus to suggest a hypothesis which is by no means a necessary one. Lehmann (1945), writing in German, uses the word 'Selbst-organisation' for it, and 'Sonderung' is also thus employed. In English it is probably best to use some phrase, such as 'localisation of regional specificity' which gives a reasonably accurate description of what is actually under discussion.

The physiological mechanisms which underlie such processes are almost completely unknown, although it is in the direction of factors resolving inherent instabilities that it seems most profitable to seek them. They offer to the embryologists of the future a problem perhaps even more important than that of the mechanisms of induction.

When we turn to consider the epigenetic organisation in later embryonic stages, we are confronted with a considerable number of different types of behaviour. In some cases we may be impressed by the dependence of an organ on the action of a specific inductor; for instance, the inductive action of the eye-cup seems to be a necessary condition for the appearance of the lens. In other cases, it is rather the independence of the developing organ which forces itself most strongly on our attention, for instance, the capacity of limb bones to develop in cultures of isolated limb buds. Throughout the whole range, however, we are really confronted with combinations, in various proportions, of the three major types of process which have already been mentioned: inductive stimuli (which in living material probably always include an element of transmitted individuation, i.e. of structural order), the localisation of regional specificity (or 'segregation' in the special sense described above), and self-individuation occurring within the developing organ.

In an organ such as the lens, the initiation of the development is, in the chick, almost completely dependent on an inductive stimulus. In this case, the natural stimulus appears to be mainly an evocation, and seems to involve little specification of the detailed structural arrangement within the lens, which arises spontaneously by a self-individuation. In the limb bud, on the other hand, the initiating inductive stimulus is probably less precisely defined, or at least is at present less definable, but again the internal organisation arises to a large extent by self-individuation; in this case we know, from the work of Saunders, that the apical ridge plays a leading part in this self-individuation, and we also have evidence of the importance of mutual interactions between various regions within the limb, as shown in the experiments of Canti & Fell on joint formation. Probably the initiation of the limb bud is brought about by a mechanism similar to that which has been more fully explored in the ear vesicle or the mesonephros. In both these, there is evidence that in normal development a complex of different inductive stimuli is at work, but the most crucial factor is a gradual localisation of regional specificity—a specificity in competence to react to the inductive stimuli by the formation of the organ in question. A similar phenomenon occurs, for instance, in the development of the mesonephros (p. 177), where a certain region of the mesoderm becomes competent to enter on this course of development, and can be caused to do so by stimuli emanating from a variety of different tissues, not all of which can be supposed to be operating in normal development. We are confronted here with a situation very similar to that which has been more fully studied in connection with the neural evocator in the amphibian gastrula. In all such cases, the problem remains of identifying the nature of the stimulus which acts during normal embryogenesis, but it is also true that one is driven to the conclusion that an even more important contribution to the whole process is the development of the appropriate competence—a development which is more or less general throughout the germ in the case of the amphibian gastrula (though the endoderm may lack it), but fairly narrowly localised in the case of the chick mesonephros. The nature of such regional localisations of competence and the factors which bring them about are still unknown, but they undoubtedly play a major part in many developmental events.

One of the next stages in embryology must be the endeavour to resolve the essential biological ideas of induction, regional localisation and individuation into their physico-chemical components. It appears unlikely that they will turn out to be dependent on any single type of biochemical or biophysical process. It may be taken as certain that the diffusion of active chemical substances plays an important role. It is probably one of the main methods by which inductive stimuli are exerted, and it is very likely that much of individuation (such as the formation of a femur of characteristic shape within a developing limb bud) is brought about by the establishment of a diffusion field of particular pattern.

It is likely also that the arising of specific self-reproducing substances or systems of substances is responsible for much of the biochemical change which must occur as the early undifferentiated cells assume their final histological character.

It is, perhaps, to similar processes that one may look for an explanation of the autonomous localisation of regional specificity; though it may be that a more profitable hypothesis would be one which envisaged a competition between mutually antagonistic autocatalytic systems. If a number of different self-reinforcing systems are set in operation throughout a given region, and if they compete with one another, perhaps for a common substrate or by some mutual stimulation or inhibition, it is easy to see that any minor initial heterogeneities may become exaggerated and developed. In this way, more complex systems of spatial order may be developed from simpler ones. Some particular mechanisms of such interactions have recently been considered by Spiegelman (1945), and others, of a more physical character, by Costello (1945). The principle has been stated above in its most general form, and we have to be prepared to find that, in different tissues, many different types of physical and chemical process are involved.

The localisation of regional specificity, which may be brought about in such ways, is a phenomenon obviously similar in some respects to certain aspects of self-individuation, by which a more or less featureless rudiment autonomously builds itself into a definite spatial configuration. We may imagine that here again the same processes, of diffusion and mutual competition, are at work and lead to the emergence of locally differentiated physical forces, which are the immediate agents moulding the tissues of the rudiment into its final form. But individuation is, of course, not always a self-individuation, since in many cases (for instance, in assimilative inductions) the morphogenesis of a mass of tissue is clearly influenced by its surroundings. In these cases, we must suppose that a field of force set up in one region is able to spread from its original site on to neighbouring material; or perhaps, since the influence is probably always mutual, it may be preferable to say that two fields of force, originally distinct from one another but of the same kind, tend to become assimilated into a single new field. A very simple example of this is seen when it is found that two neural plates, lying side by side and laterally united with one another, roll up not into a double but into a single groove. This is illustrated in the central region of the embryo shown in Fig. 40, while at the anterior end of the same embryo the coalescence is incomplete and two separate brains are developed.

Although we know almost nothing about the specific chemical substances involved in the localisation of regional differences, or of the local forces which control the morphogenesis of organs, there is one question about which a little more can tentatively be said. Can these differences be considered to make themselves felt on the cellular, or on some super-cellular level? There is no doubt that histogenesis—the development of a specific type of tissue— is a process which affects each individual cell. Even in tissues which contain a number of cell types, each cell can be individually recognised as belonging to one or other of the characteristic categories. It may seem at first sight tempting to suppose that all of morphogenesis can eventually be explained by the interaction of individual cells with one another and with their surrounding. Thus chemical changes within the cell may cause alterations in the properties of the cell surface, such that certain surfaces of contact, either with other cells, or with some external media, tend to expand or contract, in this way giving rise to forces which will deform the individual cells and, as a consequence, the whole cellular agglomeration. It can hardly be doubted that such processes play a part in the morphogenesis of many animals, particularly those in which the early embryonic stages involve rather small numbers of cells. In a form such as the chick, however, the most important morphogenetic events do not occur until the embryo contains very numerous cells; the various embryonic tissues and organ rudiments measure several cell-diameters in each direction, and the most frequent type of contact, for a cell of any given type, is with another cell of a similar type. The origin of morphogenetic forces can only be sought in places where there is contact between different elements. In multi-cellular embryos such as the chick, we must therefore seek them in interactions between cell-masses (tissues) and between such cell-masses and non-cellular media. We may use again the example of the formation of a single groove from a double neural plate: the forces involved cannot be derived directly from the properties of the individual cells, many of which will be buried within the depth of the presumptive neural material; it is rather by an interaction of the two masses of neural tissue that the unification takes place.

This argument, that the immediate seat of origin of the morphogenetic forces is often to be found in cell-masses, and not in individual cells, does not, of course, contradict the obvious fact that the character of a cell-mass is itself the result of the nature of the cells of which it is composed. The cellular nature is affected not only by the epigenetic events through which the cell has passed, but also by its intrinsic genetic constitution. Thus if the cells of the limb bud contain mutant alleles of the Creeper or Polydactyly loci, the self-individuation of the rudiment will be abnormal and will lead to the formation of the mutant morphology. The alteration in the genetic constitution of the cells has caused a change in their 'prospective potency', to use a somewhat old-fashioned terminology, so that they develop into an abnormal end-result when acted on by the normal epigenetic processes. How complex the interaction between inherent constitution and external stimuli may be is well seen in the behaviour of genetically different melanophores (p. 198), a subject which has only been touched on in this book. We have mentioned, however, that melanophores from barred breeds, when transplanted into a non-barred host, will produce barring if they reach suitable feathers, but some features of the detailed character of the barring will depend entirely on the genetic type of the melanophores, while others are influenced by regional factors in the host.

It is not unexpected that it should prove possible by means of external agents to produce alterations in the properties of a mass of cells similar to those caused by abnormalities in its genetic make-up. Such externally produced alterations are, of course, known as phenocopies, and several good examples of them have been described in the chick, particularly by Landauer and his associates (Chap. X). Investigations of phenocopy formation may be expected to throw light both on the biochemical mechanisms and on the time relations of the activities of the genes concerned. Unfortunately the interpretation of the data is never easy. There are a number of purely theoretical points to be settled, and definitions to be agreed upon, before any accurate discussion is possible. What in fact do we mean by the 'time of action of a gene'? Cell lineages of different genetic constitution are by that very fact chemically different from one another from the beginning of their life. Are we to interest ourselves in the time at which some chemical difference other than the mere substitution of one allele for another can be verified within the cells? Or with the time at which the genetic difference produces some other recognisable developmental effect, perhaps on neighbouring tissues or on the histology or morphology of the rudiment itself? Probably we shall have to distinguish a number of different meanings of the phrase 'time of action of the gene'; and these may have to take account of whether the developmental process involved is cell-autonomous or involves the mutual interactions of cells. Again, until one has reached some clear definition of exactly what process one is interested in, it is not clear that one is justified in equating the time at which an external treatment is applied with the time at which it is operative. An injected chemical, for instance, might be merely stored up until such later time as a developmental process occurred in which it could take part.

This is not the place to attempt a full discussion of the various points of theory which arise. But even a fairly crude terminology is adequate to point out what is probably the most interesting fact about the phenocopies so far described in the chick. That is, that one and the same external agent, when applied at different times in development, may produce alterations which simulate the effects of different genes. A good example is provided by the series of effects of insulin, which causes rumplessness when injected early, influences the expression of polydactyly when injected somewhat later, and later still causes micromelia and 'short upper beak'. One would certainly conclude from this, first, that the time at which the insulin begins to exert an effect is simply related to the time at which it is injected (being probably shortly afterwards), and secondly that different developmental systems are sensitive to insulin at different times. It seems more doubtful whether we can deduce that the genes in question act by a biochemical mechanism exactly the same as, or even very similar to, that set going by insulin injections; and, as has been pointed out above, it is not quite clear what we should mean if we were tempted to say that the time of injection of the insulin corresponded to the time of action of the genes, although it may be reasonable to suggest the hypothesis that some very striking phase of the gene's activity occurs at the time when the whole epigenetic system has been found to be sensitive to insulin.

It is also relevant, in any general consideration of the epigenetic organisation, to note that the expected biochemical action of insulin, and the only one for which there is as yet any definite evidence, is on the carbohydrate metabolism, and particularly the blood sugar. There are few cases in which we have definite evidence connecting the biochemistry of morphogenesis and histogenesis with that of the more fully studied energy-producing systems. Attempts to relate oxidative and glycolytic metabolism in various regions of the amphibian gastrula have, for instance, shown that, although such processes are connected with the epigenetic events, they are not directly responsible for them, since such phenomena as evocation, or the release of the evocator from an inactive condition, can occur by other means. It may well be that the relation between the immediate biochemical effects of insulin and the final phenocopy is also complex. This is strongly suggested by the recent work of Zwilling & DeBell (1950) on the teratological effects of sulphanilamide (p. 222). But even so, it can hardly be denied that the insulin-sensitive system (probably connected with carbohydrate metabolism) is united in some causal nexus with the factors determining the epigenetic events; and this whole nexus must pass through phases of relative instability ('epigenetic crises' Waddington, 1940a), which occur at different times according to the other elements of which the causal nexus is composed in different regions of the embryo.

We may pass to the consideration of an aspect of epigenetic organisation which seems to contrast rather sharply with that which has just been discussed. We have been concerned with phases of instability, in which it is easy to impair the perfection with which a certain developmental process is carried out. It is necessary now to draw attention to the fixity with which the possible types of developmental performance are laid down. The competences of later embryonic tissues often seem to be extremely narrowly delimited. A very good example is provided by the feather-forming epidermis. If a feather is plucked from an adult bird, and the papilla removed, the epidermis will grow over the site but will not, of itself, produce a new feather. It can be stimulated to do so if the dermal core of another papilla is transplanted so as to become covered by the regenerating epidermis. The tract-specific type of the feather, however, is determined, not by the inducing dermis, but by the competence of the reacting epidermis, which seems to have its developmental potentialities (for feather formation) limited to the production of only one particular type of feather. It was pointed out earlier (p. 195) that in most other cases in which the character of an induction is dependent on the competence of the reacting material, this material differs from the inductor in its genetic nature (as, for instance, in the induction of balancers from urodele ectoderm by anuran mesoderm), whereas here the difference can only have arisen epigenetically. The matter is mentioned again here, not because anything material can be added to the earlier discussion, but because the unravelling of the conditions which underlie the sharply contrasted qualities of embryonic tissues, so plastic in some respects and so fixed in others, seems to be one of the key problems facing epigenetics in the immediate future.

BIBLIOGRAPHY

ABERCROMBIE, M. (1937). The behaviour of epiblast grafts beneath the primitive streak of the chick. *J. Exp. Zool.* **14**, 302.

——(1939). Evocation in the chick. *Nature*, **144**, 1091.

——(1950). The effects of anterio-posterior reversal of lengths of the primitive streak in the chick. *Phil. Trans. Roy. Soc.* B, **234**, 317.

ABERCROMBIE, M. & BELLAIRS, M. R. (1951). Unpublished.

ABERCROMBIE, M. & CAUSEY, G. (1950). Identification of transplanted tissues in chick embryos by marking with phosphorus-32. *Nature*, **166**, 229.

ABERCOMBIE, M. & WADDINGTON, C. H. (1937). The behaviour of grafts of primitive streak beneath the primitive streak of the chick. *J. Exp. Biol.* **14**, 319.

ACKERMAN, W. W. & TAYLOR, A. (1948). Application of a metabolic inhibitor to the developing chick embryo. *Proc. Soc. Exp. Biol. Med.* **67**, 449.

ALEXANDER, L. E. (1937). An experimental study of the role of optic cup and overlying ectoderm in lens formation in the chick embryo. *J. Exp. Zool.* **75**, 41.

ALIBERTI, G. (1933). Fenomeni interessanti il sacco vitellino degli uccelli sopravviventi all' embrione. *Bol. Mus. Lab. Zool. Anat. Comp. Genova*, **13**.

AMPRINO, R. (1943). Correlazione quantitative fra centri nervosi e territori d' innervazione periferica durante lo sviluppo. Ricerche sperimentali sul ganglio ciliare del pollo. *Arch. Ital. Anat. Embr.* **49**, 1.

—— (1949). Ricerche sperimentali sulla morfogenesi del cristallino nell' embrione di pollo. Induzione e rigenerazione. *Arch. f. Entw. mech. Org.* **144**, 171.

—— (1950). Les conditions qui réglent l'organogenèse et la croissance des muscles. Recherches expérimentales sur les muscles extrinsèques de l'oeil du poulet. *Acta Anat. (Basel)*, **10**, 38.

ANCEL, P. (1945*a*). L'achondroplasie, sa réalisation expérimentale—sa pathogénie. *Ann. d' Endocrinol.* **6**, 1.

—— (1945*b*). Les variations individuelles dans les expériences de tératogenèse. *La Revue scientif.* **83**, 99.

—— (1947). Recherches sur la réalisation expérimentale de la célosomie. *Arch. Anat., Strasbourg*, **30**, 1.

ANCEL, P. & LALLEMAND, S. (1940). Sur la spécificité de l'action tératogène de la colchicine chez l'embryon du poulet. *C.R. Acad. Sci.* **210**, 710.

—— —— (1942). Sur une malformation du bec et des membres obtenue chez l'embryon du poulet à l'aide de sulphamides. *C.R. Soc. Biol.* **136**, 255.

—— —— (1944). Sur les malformations du bec, spontanées et provoquées, chez l'embryon du poulet. *C.R. Soc. Biol.* **138**, 374.

—— —— (1941). Sur l'obtention de la strophosomie chez le poulet à l'aide de la ricine. *C.R. Acad. Sci.* **212**, 312.

ASSHETON, R. (1896). An experimental examination into the growth of the blastoderm of the chick. *Proc. Roy. Soc.* B, **60**, 349.

BALFOUR, F. M. (1873). The development and growth of the layers of the blastoderm. *Q.J. micr. Sci.* **13**, 266.

BARFURTH, D. & DRAGENDORFF, O. (1902). Versuche über Regeneration des Auges und der Linse beim Hühnerembryo. *Anat. Anz.* **21**, 185.

BARNETT, S. A. & BOURNE, G. (1941). The distribution of ascorbic acid (vitamin C) in the early stages of the developing chick embryo. *J. Anat.* **75**, 251.

—— —— (1942). Distribution of ascorbic acid (vitamin C) in cells and tissues of the developing chick. *Q.J. micr. Sci.* **83**, 259.

BARRON, D. H. (1943). The early development of the motor cells and columns in the spinal cord of the sheep. *J. Comp. Neurol.* **78**, 1.

—— (1946). Observations on the early differentiation of the motor neuroblasts in the spinal cord of the chicken. *J. Comp. Neurol.* **85**, 149.

—— (1947). Some effects of amputation of the chick wing bud on the early differentiation of the associated motor neuroblasts. *Anat. Rec.* **97**, 320.

—— (1948). Some effects of amputation of the chick wing bud on the early differentiation of the motor neuroblasts in the associated segments of the spinal cord. *J. Comp. Neurol.* **88**, 93.

BARTELMEZ, G. W. (1918). The relation of the embryo to the principal axis of symmetry in the bird's egg. *Biol. Bull.* **35**, 319.

BATESON, W. (1902). Experiments with poultry. *Rep. Evol. Ctte. Roy. Soc.* **1**, 87.

BAUMANN, L. & LANDAUER, W. (1943). Polydactyly and anterior horn cells in the fowl. *J. Comp. Neurol.* **79**, 153.

—— —— (1944). On the expression of polydactylism in the wings of fowl. *Anat. Rec.* **90**, 225.

BAUTZMANN, H. (1932). Experimentelle Analyse des organisatorischen Geschehens in der Primitiventwicklung von Amphibien. Determinationszustand und Aufgabenverteilung der Randzonenanlagen im Organisationsprozess. *Anat. Anz.* **75**, 220.

BAUTZMANN, H., HOLTFRETER, J., SPEMANN, H. & MANGOLD, O. 1932. Versuche zur Analyse der Induktionsmittel in der Embryonalentwicklung. *Naturwiss.* **20**, 971.

DE BEER, G. R. (1940). *Embryos and Ancestors.* Oxford Univ. Press.

BEQUELIN. (1751). Mémoire sur l'art de couver les œufs ouverts. *Hist. de l'Ac. Roy. des Sci. et Bell. Lett. de Berlin,* **5**, 71.

BIJTEL, H. M. (1931). Über die Entwicklung des Schwanzes bei Amphibien. *Arch. f. Entw. mech.* **125**, 448.

—— (1936). Die Mesodermbildungspotenzen der hinteren Medullarplattenbezirke bei *Amblystoma mexicanum* in Bezug auf Schwanzbildung. *Arch. f. Entw. mech. Org.* **134**, 262.

BLIVAISS, B. B. (1947). Interrelations of thyroid and gonad in the development of plumage and other sex characters in Brown Leghorn roosters. *Physiol. Zool.* **20**, 67.

BOND, C. J. (1926). Further observations on polydactyly and heterodactyly in fowls. *J. Genet.* **16**, 253.

—— (1932). On the genetic significance of hemilateral asymmetry in the vertebrate organism. William Wittening Lecture, University of Birmingham.

BOYDEN, E. A. (1924). An experimental study of the development of the avian cloaca. *J. Exp. Zool.* **40**, 437.

—— (1927). Experimental obstruction of the mesonephric ducts. *Proc. Soc. Exp. Biol. Med.* **24**, 572.

BRACHET, A. (1912). Développement in vitro de blastomes et de jeunes embryons de Mammifères. *C.R. Acad. Sci.* **155**, 1191.

—— (1913). Recherches sur le déterminisme héréditaire de l'œuf des Mammifères. Développement in vitro de jeunes vésicules blastodermiques du lapin. *Arch. Biol.* **28**, 447.

—— (1935). *Traité d'Embryologie des Vertébrés.* 2nd edition revised by A. Dalcq and P. Gerard. Paris.

BRACHET, J. (1940). Étude histochimique des Protéines au cours du développement embryonnaire des Poissons, des Amphibiens et des Oiseaux. *Arch. Biol.* **51**, 167.

—— (1944). *Embryologie chimique.* Paris: Masson; Liége: Desoer.

—— (1949). *Unités biologiques douées de continuité génétique.* Paris.

BRAUS, H. (1908). Entwicklungsgeschichtliche Analyse der Hyperdaktylie. *Münch. Med. Wochenschr.* **55**, 386.

BREMER, J. L. (1928). Experiments on the aortic arches in the chick. *Anat. Rec.* **37**, 225.

BUCHANAN, J. W. (1926). Regional differences in rate of oxidations in the chick blastoderm as shown by susceptibility to hydrocyanic acid. *J. Exp. Zool.* **45**, 141.

BUDDINGH, G. J. & POLK, A. D. (1939). Experimental meningococcus infection of the chick embryo. *J. Exp. Med.* **70**, 485.

BUEKER, E. D. (1943). Intracentral and peripheral factors in the differentiation of motor neurons in transplanted lumbo-sacral spinal cords of chick embryos. *J. Exp. Zool.* **93**, 99.

—— (1945). The influence of a growing limb on the differentiation of somatic motor neurons in transplanted avian spinal cord segments. *J. Comp. Neurol.* **82**, 335.

—— (1947). Limb ablation experiments on the embryonic chick and its effects as observed on the mature nervous system. *Anat. Rec.* **97**, 157.

—— (1948). Implantation of tumors in the hind-limb field of the embryonic chick and the developmental response of the lumbo-sacral nervous system. *Anat. Rec.* **102**, 369.

BUEKER, E. D. & BELKIN, M. (1947). Growth of the lumbo-sacral nervous system of the chick after the substitution of tumours for the limb periphery. *Anat. Rec.* **97**, 323.

BURKE, V., SULLIVAN, N. P., PETERSEN, H. & WEED, R. (1944). Ontogenetic change in the antigenic specificity of the organs of the chick. *J. Infect. Dis.* **74**, 225.

BURR, H. S. & HOVLAND, C. I. (1937). Bioelectrical potential gradients in the chick. *Yale J. Biol. Med.* **9**, 247.

BUTLER, E. (1935). The developmental capacity of regions of the unincubated chick blastoderms as tested in chorio-allantoic grafts. *J. Exp. Zool.* **70**, 357.

BYERLY, T. C. (1926). Studies in growth. II. Local growth in 'dead' embryos. *Anat. Rec.* **32**, 249.

—— (1932). Growth of the chick embryo in relation to its food supply. *J. Exp. Biol.* **9**, 15.

BYERLY, T. C., TITUS, H. W., ELLIS, N. R. & LANDAUER, W. (1935). A new nutritional disease of the chick embryo. *Proc. Soc. Exp. Biol. Med.* **32**, 1542.

BYTINSKI-SALZ, H. (1929). Untersuchungen über das Verhalten des presumptiven Gastrulaektoderms der Amphibien bei heteroplastischer und xenoplastischer Transplantation ins Gastrocoel. *Arch. f. Entw. mech. Org.* **114**, 594.

CAIN, A. J. (1950). The histochemistry of lipoids in animals. *Biol. Rev.* **25**, 73.

CAIRNS, J. M. (1937). The development of Hensen's node when grafted directly to the chick blastoderm. M.S. Thesis, Univ. Rochester. (Quoted Rudnick, 1944.)

—— (1941). The 'early lethal' action of the homozygous Creeper factor in the chick. *J. Exp. Zool.* **88**, 481.

CAIRNS, J. M. & GAYER, K. (1943). Identification of chick embryos homozygous for the Creeper factor. *J. Exp. Zool.* **92**, 229.

VAN CAMPENHOUT, E. (1946). Le laboratoire d'histo-embryologie de l'Université de Louvain (1936–1946). *Arch. Med. Belg.* **1**, 260.

CARPENTER, E. (1942). Differentiation of chick embryo thyroids in tissue culture. *J. Exp. Zool.* **89**, 407.

TEN CATE, G. (1949). The presence of adult lens protein in young lens vesicles of chicken and frog embryos. *Comm. journees cyto-emb. belgo-neerl. Gand*, p. 92.

TEN CATE, G. & VAN DOORENMAALEN, W. J. (1950). Analysis of the development of the eye-lens in chicken and frog embryos by means of the precipitin reaction. *Proc. Acad. Sci. Amst.* **53**, 6.

CHANG, T. K. (1939). The development of polydactylism in a special strain of *Mus musculus*. *Peiping Nat. Hist. Bull.* **14**, 119.

CHARLES, D. R. & RAWLES, M. E. (1940). Tyrosinase in feather germs. *Proc. Soc. Exp. Biol. Med.* **43**, 55.

CHILD, C. M. (1946). Organisers in development and the organiser concept. *Physiol. Zool.* **19**, 89.

CLARKE, L. F. (1936). Regional differences in eye-forming capacity of the early chick blastoderm as studied in chorio-allantoic grafts. *Physiol. Zool.* **9**, 102.

COLE, R. K. (1942). The 'talpid lethal' in the domestic fowl. *J. Hered.* **33**, 83.

CONRAD, R. M. & SCOTT, H. M. (1938). The formation of the egg of the domestic fowl. *Physiol. Rev.* **18**, 481.

COSTELLO, D. P. (1945). Segregation of oöplasmic constituents. *J. El. Mitch. Sci. Soc.* **61**, 277.

CRAVENS, W. W., McGIBBON, W. H. & SEBESTA, E. J. (1944). Effect of biotin deficiency on embryonic development in the domestic fowl. *Anat. Rec.* **90**, 55.

CREW, F. A. E. & MUNRO, S. S. (1938). Gynandromorphism and lateral asymmetry in the fowl. *Proc. Roy. Soc. Edin.* **58**, 114.

—— —— (1939). Lateral asymmetry in the fowl. *Proc. VIIth World's Poultry Congress*, p. 61.

CRUICKSHANK, E. M. (1931). The effect of diet on the chemical composition, nutritive value and hatchability of the egg. *Nutr. Absts. & Revs.* **10**, 645.

DALCQ, A. (1933). La determination de la vesicule auditive chez le Discoglosse. *Arch. anat. micr.* **29**, 389.

—— (1937). Les plans d'ébauches chez les Vertébrés et la signification morphologique des annexes fœtales. *Ann. Soc. Roy. Zool. belg.* **68**, 69.

—— (1938). Étude micrographique et quantitative de la Mérogonie double chez *Ascidiella scrabra. Arch. Biol.* **49**, 399.

—— (1941). *L'Œuf et son dynanisme organisateur.* Paris.

—— (1947). Recent experimental contributions to brain morphogenesis in amphibians. *Growth Symp.* **6**, 85.

DALCQ, A. & HUANG, A. C. (1948). Effets de la division, par ligature, de la blastula et de la gastrula du Triton. *C.R. Soc. Biol.* **142**, 1312.

DALTON, A. J. (1935). The potencies of portions of young chick blastoderms as tested in chorio-allantoic grafts. *J. Exp. Zool.* **71**, 17.

DANCHAKOFF, V. (1924). Wachstum transplantierter embryonaler Gewebe in der Allantois. *Zeits. Anat.* **74**, 401.

—— (1926). Lens ectoderm and optic vesicles in allantois grafts. *Contrib. Embryol. Carneg. Instn,* **18**, 63.

DANCHAKOFF, V. & GAGARIN, A. (1929). Embryoherz in der Chorio-Allantois des Hühnchens. *Zeits. Anat.* **89**, 754.

DANFORTH, C. H. (1929). The effect of foreign skin on feather pattern in the common fowl. *Arch. f. Entw. mech. Org.* **115**, 242.

DANFORTH, C. H. & FOSTER, F. (1929). Skin transplantation as a means of studying genetic and endocrine factors in fowl. *J. Exp. Zool.* **92**, 443.

DANFORTH, C. H. (1932–3). Artificial and hereditary suppression of the sacral vertebrae in the fowl. *Proc. Soc. Exp. Biol. Med.,* **30**, 143.

—— (1939). Direct control of avian color pattern by the pigmentoblasts. *J. Hered.* **30**, 173.

DAVID, P. R. (1936). Studies on the Creeper fowl. X. A study of the mode of action of a lethal factor by explantation methods. *Arch. f. Entw. mech.* **135**, 521.

DAVIS, J. O. (1944). Photochemical spectral analysis of neural tube formation. *Biol. Bull.* **87**, 73.

DEMUTH, F. (1939). Über die Beziehung des Energiestoffwechsels zu Wachstum und Differenzierung. II. Die Wirkung von Sauerstoff auf Wachstum und Entwicklung von Hühnerembryonen. *Proc. Acad. Sci. Amst.* **42**, 506.

DERRICK, G. E. (1937). An analysis of the early development of the chick by means of the mitotic index. *J. Morph.* **61**, 257.

VAN DETH, J. H. M. G. (1939). Lensinductie en lensregeneratie bij het kippenembryo. *Acad. Proefschr. Amsterdam, 6th July.*

—— (1940). Induction et régénération du cristallin chez l'embryon de la poule. *Acta neerl. Morph.* **3**, 151.

DETWILER, S. R. (1936). *Neuro-embryology.* New York: Macmillan Co.

DORRIS, F. (1936). Differentiation of pigment cells in tissue cultures of chick neural crest. *Proc. Soc. Exp. Biol. Med.* **34**, 448.

—— (1938a). Differentiation of the chick eye *in vitro*. *J. Exp. Zool.* **78**, 386.

—— (1938b). The production of pigment *in vitro* by chick neural crest. *Arch. f. Entw. mech. Org.* **138**, 321.

—— (1939). The production of pigment by chick neural crest in grafts to the 3-day limb bud. *J. Exp. Zool.* **80**, 315.

DRAGENDORFF, O. (1903). Experimentelle Untersuchungen über Regenerationsvorgänge am Auge und an der Linse von Hühnerembryonen. *Inaug. Diss. Rostock.* (Quoted by Mangold, 1931.)

DRAGOMIROV, N. (1931). Ein Fall abnormer Augenbildung beim Entenembryo. *Arch. f. Entw. mech. Org.* **125**, 189.

DUNN, L. C. (1925). The inheritance of rumplessness in the domestic fowl. *J. Hered.* **16**, 127.

DUNN, L. C. & LANDAUER, W. (1925). Two types of rumplessness in domestic fowls. *J. Hered.* **16**, 151.

—— —— (1927). The lethal nature of the Creeper variation in the domestic fowl. *Amer. Nat.* **60**, 574.

—— —— (1934). The genetics of the rumpless fowl with evidence of a case of changing dominance. *J. Genet.* **29**, 217.

DU SHANE, G. P. (1935). An experimental study of the origin of pigment cells in Amphibia. *J. Exp. Zool.* **72**, 1.

—— (1944). The embryology of vertebrate pigment cells. II. Birds. *Quart. Rev. Biol.* **19**, 98.

—— (1948). The biology of Melanomas. *Spec. Publ. N.Y. Acad. Sci.* **4**, 1.

DU TOIT, P. J. (1913). Untersuchungen über das Synsacrum und den Schwanz von *Gallus domestica*. *Jena. Zeits. Naturwiss.* **49**, 1.

DUVAL, M. (1884). De la formation du blastoderme dans l'œuf d'oiseau. *Ann. Sci. Nat. Zool.*, 6th ser., **18**, 1.

—— (1888). *Atlas d'embryologie*. Paris.

EASTLICK, J. L. (1939a). The point of origin of the melanophores in chick embryos as shown by means of limb bud transplants. *J. Exp. Zool.* **82**, 131.

—— (1939b). The role of heredity versus environment in limb bud transplants between different breeds of fowl. *Science*, **17**, 181.

—— (1939c). Reciprocal heterotransplantation of limb primordia between duck, turkey, guinea fowl and chick embryos. *Nature*, **144**, 380.

—— (1940). The localisation of pigment forming areas in the chick blastoderm at the primitive streak stage. *Physiol. Zool.* **13**, 202.

—— (1941). Manifestations of incompatibility in limb grafts made between bird embryos of different species. *Physiol. Zool.* **14**, 136.

—— (1943). Studies on the transplanted embryonic limbs of the chick. I. The development of muscle in nerveless and in innervated grafts. *J. Exp. Zool.* **93**, 27.

EASTLICK, J. L. & WORTHAM, R. A. (1947). Studies on transplanted limbs of the chick. III. The replacement of muscle by adipose tissue. *J. Morph.* **80**, 369.

EBERT, J. D. (1949). An analysis of the effects of certain anti-organ sera on the development of the early chick blastoderm *in vitro*. *Anat. Rec.* **105**, 12.

—— (1950). An analysis of the effects of anti-organ sera on the development, *in vitro*, of the early chick blastoderm. *J. Exp. Zool.* **115**, 351.

'ESPINASSE, P. G. (1939). The developmental anatomy of the Brown Leghorn breast feather and its reactions to oestrone. *Proc. Zool. Soc. Lond.* A, **109**, 247.

EVANS, J. H. (1943). The independent differentiation of the sensory areas of the avian inner ear. *Biol. Bull.* **84**, 252.

FELL, H. B. (1928). The development *in vitro* of the isolated otocyst of the embryonic fowl. *Arch. f. Exp. Zellforsch.* **7**, 69.

—— (1932). The osteogenic capacity *in vitro* of periosteum and endosteum isolated from the limb skeleton of fowl embryos and young chicks. *J. Anat.* **66**, 157.

—— (1933). Chondrogenesis in cultures of endosteum. *Proc. Roy. Soc.* B, **112**, 417.

—— (1939). The origin and developmental mechanics of the avian sternum. *Phil. Trans. Roy. Soc.* B, **229**, 407.

FELL, H. B. & CANTI, R. G. (1934). Experiments on the development *in vitro* of the avian knee-joint. *Proc. Roy. Soc.* B, **116**, 316.

FELL, H. B. & LANDAUER, W. (1935). Experiments on skeletal growth and development *in vitro* in relation to the problem of avian phokomelia. *Proc. Roy. Soc.* B, **118**, 133.

FELL, H. B. & ROBISON, R. (1929). The growth, development and phosphatase activity of embryonic avian femora and limb buds cultivated *in vitro*. *Biochem. J.* **23**, 767.

—— —— (1930). The development and phosphatase activity *in vivo* and *in vitro* of the mandibular skeletal tissue of the embryonic fowl. *Biochem. J.* **24**, 1905.

—— —— (1934). The development of the calcifying mechanism in avian cartilage and osteoid tissue. *Biochem. J.* **28**, 2243.

FÉRE, C. (1895). Note sur le sort de blastodermes de poulet implantés dans les tissues d'animaux de la même espèce. *C.R. Soc. Biol.* **47**, 331.

FÉRE, C. & ELIAS, N. (1897). Note sur l'évolution d'organes d'embryons de poulet greffés sous la peau d'oiseaux adultes. *Arch. anat. micr.* **1**, 417.

FERRET, P. & WEBER, A. (1904). Malformations du système nerveux central de l'embryon de poulet obtenu expérimentalement. *C.R. Soc. Biol.* **56**, 187, 286.

FILOGAMO, G. (1950). Consequenze della demolizione dell' abbozzo dell' occhio sullo sviluppo del lobo ottico nell embrione di pollo. *Riv. di Biol.* **42**, 73.

FOULKS, J. G. (1943). An analysis of the source of melanophores in regenerating feathers. *Physiol. Zool.* **16**, 351.

FOX, M. H. (1949). Analysis of some phases of melanoblast migration in the Barred Plymouth Rock embryos. *Physiol. Zool.* **22**, 1.

FRANKE, K. W., MOXON, A. L., POLEY, W. E. & TULLY, W. C. (1936). Monstrosities produced by the injection of selenium salts into hen's eggs. *Anat. Rec.* **65**, 19.

BIBLIOGRAPHY 247

FRAPS, R. M. & JUHN, M. (1936a). Developmental analysis in plumage. II. Plumage configurations and the mechanism of feather development. *Physiol. Zool.* **9**, 319.

—— —— (1936b). Developmental analysis in plumage. III. Field functions in the breast tracts. *Physiol. Zool.* **9**, 378.

FRÖHLICH, K. (1936). Experimentelle Erzeugung von Sirenen beim Hühnchen und Bemerkungen zur sekundären Körperentwicklung. *Arch. f. Entw. mech. Org.* **134**, 348.

FUGO, N.W. (1940). Effect of hypophysectomy in the chick embryo. *J. Exp. Zool.* **85**, 271.

GABRIEL, M. L. (1946a) The effect of local applications of colchicine on leghorn and polydactylous chick embryos. *J. Exp. Zool.* **101**, 339.

—— (1946b). Production of strophosomy in the chick embryo by local applications of colchicine. *J. Exp. Zool.* **101**, 351.

GAERTNER, R. A. (1949). Development of the posterior trunk and tail of the chick embryo. *J. Exp. Zool.* **111**, 157.

GALLERA, J. (1947). Effets de la suspension précoce de l'induction neurale sur la partie préchordale de la plaque neurale chez les Amphibiens. *Arch. Biol.* **58**, 221.

GALLERA, J. & OPRECHT, E. (1948). Sur la distribution des substances basophiles cytoplasmiques dans le blastoderme de la poule. *Rev. suisse Zool.* **55**, 243.

GAYER, K. (1942). A study of coloboma and other abnormalities in transplants of eye primordia from normal and Creeper chick embryos. *J. Exp. Zool.* **89**, 103.

GAYER, K. & HAMBURGER, V. (1943). The developmental potencies of eye primordia of homozygous Creeper chick embryos tested by orthotopic transplantation. *J. Exp. Zool.* **93**, 147.

GILLMAN, J., GILBERT, C. & GILLMAN, T. (1948). A preliminary report on Hydrocephalus, Spina Bifida and other congenital anomalies in the rat produced by Trypan Blue. *S. Afr. J. Med. Sci.* **13**, 47.

GLASER, O. C. (1914). On the mechanism of morphological differentiation in the nervous system. I. The transformation of a neural plate into a neural tube. *Anat. Rec.* **8**, 525.

GLUECKSMANN, A. (1951). Cell deaths in normal vertebrate ontogeny. *Biol. Rev.* **26**, 59.

GLUECKSOHN-SCHOENHEIMER, S. (1949). The effects of a lethal gene responsible for duplications and twinning in mouse embryos. *J. Exp. Zool.* **110**, 47.

GRÄPER, L. (1907). Untersuchungen über die Herzbildung der Vögel. *Arch. f. Entw. mech. Org.* **24**, 375.

—— (1929a). Die Primitiventwicklung des Hühnchens nach stereokinematographischen Untersuchungen, kontrolliert durch vitale Farbmarkierung und verglichen mit der Entwicklung anderer Wirbeltiere. *Arch. f. Entw. mech. Org.* **116**, 382.

—— (1929b). Die Primitiventwicklung des Hühnchens, verglichen mit der anderer Wirbeltiere. *Verh. Anat. Ges.* **67**, 90.

—— (1930a). Zur Gastrulation der Wirbeltiere. *Anat. Anz.* **69**, 248.

GRÄPER, L. (1930b). Wachstumsvorgänge beobachtet mittelst Stereokomparation von Reihenaufnahmen lebender Hühnerembryonen. *Zeits. Anat. Entwges.* **92**, 700.

—— (1931). Primitiventwicklung und einheitliche Erklärung von Doppelbildungen. *Verh. Anat. Ges.* **10**, 35.

—— (1932). Zur Entwicklung der hinteren Körperhälfte des Hünchens. *Anat. Anz.* **75**, 200.

—— (1933). Beitrag zur Frage der sekundären Körperentwicklung und der hinteren Extremitäten beim Hünchen. *Arch. f. Entw. mech. Org.* **128**, 766.

GRAY, P. (1939). Experiments with direct currents on chick embryos. *Arch. f. Entw. mech. Org.* **139**, 732.

GRAY, P. & WORTHING, H. (1941). Experiments on chemical interference with the early morphogenesis of the chick. I. The effects of tetanus toxin on the morphogenesis of the central nervous system. *J. Exp. Zool.* **86**, 423.

GREENWOOD, A. W. & BLYTHE, J. S. S. (1951). Genetic and somatic aberrations in two asymmetrically marked fowls from sex-linked crosses. *Heredity*, **5**, 215.

GREGORY, P. W., ASMUNDSON, V. S., GOSS, H. & LANDAUER, W. (1939). Glutathione values of Cornish lethal and Creeper embryos compared with normal sibs. *Growth*, **3**, 75.

GRODZIŃSKI, Z. (1933). Über die Entwicklung von unterkühlten Hühnereiern. *Arch. f. Entw. mech. Org.* **129**, 512.

—— (1934a). Zur Kenntnis der Wachstumvorgänge der Area vasculosa beim Hünchen. *Bull. int. Acad. Cracovie, Série B* (II), 415.

—— (1934b). Weitere Untersuchungen über den Einfluss der Unterkühlung auf die Entwicklung der Hühnereier. *Arch. f. Entw. mech. Org.* **131**, 653.

—— (1935). Die Entwicklung der Venen in der Keimscheibe des Hühnchens. *Bull. int. Acad. Cracovie, Série B* (II), 305.

GRÜNWALD, P. (1935). Teratologische Untersuchungen über die mutmässlichen Beziehungen der abnormen und normalen Medullaranlage zur Entwicklung der Urwirbel beim Huhn. *Arch. f. Entw. mech. Org.* **133**, 664.

—— (1936a). Zur Entwicklungsmechanik des Urogenitalsystems beim Huhn. *Arch. f. Entw. mech. Org.* **136**, 786.

—— (1936b). Experimentelle Untersuchungen uber die Beziehungen der Medullaranlage zur Entwicklung der Urwirbel beim Huhn. *Arch. f. Entw. mech. Org.* **135**, 389.

—— (1937). Ein Fall von omphalocephalus syncephalis bei der Ente. *Zeits. Anat. Entwges.* **107**, 782.

—— (1941). Normal and abnormal detachment of body and gut from the blastoderm in chick embryos, with remarks on the early development of the allantois. *J. Morph.* **69**, 83.

—— (1942). Experiments on the distribution of the nephrogenic potency in the embryonic mesenchyme. *Physiol. Zool.* **15**, 396.

—— (1943). Stimulation of nephrogenic tissue by normal and abnormal inductors. *Anat. Rec.* **86**, 321.

—— (1947a). Mechanisms of abnormal development, I, II, III. *Arch. Pathol.* **44**, 398, 495, 648.

GRÜNWALD, P. (1947b). Studies on developmental pathology. IV. The development of malformations with median defects of the caudal part of the body. *J. Morph.* **81**, 97.

—— (1947c). Deviation of axial organs as a cause of sirenoid malformations. *Anat. Rec.* **97**, 339.

GURWITSCH, A. (1922). Über den Begriff des embryonalen Feldes. *Arch. f. Entw. mech. Org.* **51**, 383.

HAMBURGER, V. (1934). The effects of wing bud extirpation on the development of the central nervous system in chick embryos. *J. Exp. Zool.* **68**, 449.

—— (1938). Morphogenetic and axial self-differentiation of transplanted limb primordia of 2-day chick embryos. *J. Exp. Zool.* **77**, 379.

—— (1939a). Motor and sensory hyperplasia following limb bud transplantations in chick embryos. *Physiol. Zool.* **12**, 268.

—— (1939b). The development and innervation of transplanted limb primordia of chick embryos. *J. Exp. Zool.* **80**, 347.

—— (1939c). A study of hereditary chrondrodystrophia in the chick (Creeper fowl) by means of embryonic transplantation. *Proc. Soc. Exp. Biol. Med.* **41**, 13.

—— (1939d). Correlations between nervous and non-nervous structures during development. *Coll. Net.* **14**.

—— (1941). Transplantation of limb primordia of homozygous and heterozygous chondrodystrophic ('Creeper') chick embryos. *Physiol. Zool.* **14**, 355.

—— (1942a). The developmental mechanics of hereditary abnormalities in the chick. *Biol. Symp.* **6**, 311.

—— (1942b). *A manual of experimental embryology.* Univ. Chicago Press.

—— (1946). Isolation of the brachial segments of the spinal cord of the chick embryo by means of tantalum foil blocks. *J. Exp. Zool.* **103**, 113.

—— (1948). The mitotic patterns in the spinal cord of the chick embryo and their relation to histogenetic processes. *J. Comp. Neurol.* **88**, 221.

HAMBURGER, V. & HABEL, K. (1948). Teratogenic and lethal effects of Influenza-A and Mumps viruses on early chick embryos. *Proc. Soc. Exp. Biol. Med.* **66**, 608.

HAMBURGER, V. & HAMILTON, H. L. (1951). A series of normal stages in the development of the chick embryo. *J. Morph.* **88**, 49.

HAMBURGER, V. & KEEFE, E. (1944). The effects of peripheral factors on the proliferation and differentiation in the spinal cord of chick embryos. *J. Exp. Zool.* **96**, 223.

HAMBURGER, V. & LEVI-MONTALCINI, R. (1949). Proliferation, differentiation and degeneration in the spinal ganglia of the chick embryo under normal and experimental conditions. *J. Exp. Zool.* **111**, 457.

HAMBURGER, V. & WAUGH, M. (1940). The primary development of the skeleton in nerveless and poorly innervated limb transplants of chick embryos. *Physiol. Zool.* **13**, 367.

HAMILTON, H. L. (1940). A study of the physiological properties of melanophores with special reference to their role in feather coloration. *Anat. Rec.* **78**, 525.

HAMMOND, W. S. (1949). Formation of the sympathetic nervous system in the trunk of the chick embryo following removal of the thoracic neural tube. *J. Comp. Neurol.* **91**, 67.

250   BIBLIOGRAPHY

HAMMOND, W. S. & YNTENA, C. L. (1947). Depletion in the thoraco-lumbar sympathetic system following removal of the neural crest in the chick. *J. Comp. Neurol.* **86**, 237.

HARMAN, M. T. & NELSON, F. (1941). Polydactyl feet of two strains of chicks. *Amer. Nat.* **75**, 540.

HATT, P. (1934). L'induction d'une plaque médullaire secondaire chez le Triton par l'implantation d'un morceau de ligne primitive de poulet. *Arch. anat. micr.* **30**, 131.

HEATLEY, N. G. & LINDAHL, P. E. (1937). The distribution and nature of glycogen in the amphibian embryo. *Proc. Roy. Soc.* B, **122**, 395.

HILLEMANN, H. H. (1943). An experimental study of the development of the pituitary gland in chick embryos. *J. Exp. Zool.* **93**, 347.

HINRICHS, M. A. (1927). Modification of development on the basis of differential susceptibility to radiation. IV. Chick embryos and ultraviolet radiation. *J. Exp. Zool.* **47**, 309.

—— (1931). Ultraviolet point radiation in production of developmental abnormalities in the chick embryo. *Proc. Soc. Exp. Biol. Med.* **28**, 1059.

HOADLEY, L. (1924). The independent differentiation of isolated chick primordia in chorio-allantoic grafts. I. The eye, nasal region and mesencephalon. *Biol. Bull.* **46**, 281.

—— (1925a). The differentiation of isolated chick primordia in chorio-allantoic grafts. II. The effect of the presence of the spinal cord on the differentiation of the somitic region. *J. Exp. Zool.* **42**, 143.

—— (1925b). The differentiation of isolated chick primordia in chorio-allantoic grafts. III. On the specificity of nerve processes arising from the mesencephalon in grafts. *J. Exp. Zool.* **42**, 163.

—— (1926a). Developmental potencies of parts of the early blastoderm of the chick. I, II and III. *J. Exp. Zool.* **43**, 151.

—— (1926b). The in situ development of sectioned chick blastoderm. *Arch. de Biol.* **36**, 225.

—— (1927). Concerning the organisation of potential areas in the chick blastoderm. *J. Exp. Zool.* **48**, 459.

—— (1929). Differentiation versus cleavage in chorio-allantoic grafts. *Arch. f. Entw. mech. Org.* **116**, 278.

HOBSON, L. B. (1941). On the ultra structure of the neural plate and tube of the early chick embryo, with notes on the effects of dehydration. *J. Exp. Zool.* **88**, 107.

HOLMDAHL, D. E. (1924–5). Experimentelle Untersuchungen über die Lage der Grenze zwischen primärer und sekundärer Körperentwicklung beim Huhn. *Anat. Anz.* **59**, 393.

—— (1925–6). Die erste Entwicklung des Körpers bei den Vögelen und Säugetieren. I–V. *Morph. Jahrb.* **54**, 333; **55**, 112.

—— (1933). Die zweifache Bildungsweise des zentralen Nervensystems bei den Wirbeltieren. *Arch. f. Entw. mech. Org.* **129**, 206.

—— (1935). Primitivstreifen bzw. Rumpfschwanzknospe im Verhältnis zur Körperentwicklung. *Zeits. mikr.-anat. Forsch.* **38**, 409.

—— (1936). Neue Gesichtspunkte zur frühen Embryonalentwicklung. *Uppsala Läk. för. forh. Ny folyd.* **42**, 1.

HOLMDAHL, D. E. (1939a). Die Morphogenese des Vertebratenorganismus vom formalen und experimentellen Gesichtspunkt. *Arch. f. Entw. mech. Org.* **139**, 191.

—— (1939b). Die formalen Verhältnisse während der Entwicklung der Rumpfschwanzknospe beim Huhn. *Anat. Anz.* **88**, 127.

HOLMES, A. (1935). The pattern and symmetry of adult plumage units in relation to the order and locus of origin of the embryonic feather papillae. *Amer. J. Anat.* **56**, 513.

HOLTFRETER, J. (1933). Der Einfluss von Wirtsalter und verschiedenen Organbezirken auf die Differenzierung von angelagertem Gastrulaektoderm. *Arch. f. Entw. mech. Org.* **127**, 619.

—— (1938a). Differnzierungspotenzen isolierter Teile der Anurengastrula. *Arch. f. Entw. mech. Org.* **138**, 657.

—— (1938b). Differenzierungspotenzen isolierter Teile der Urodelengastrula. *Arch. f. Entw. mech. Org.* **138**, 522.

—— (1943). A study of the mechanics of gastrulation. I. *J. Exp. Zool.* **94**, 261.

—— (1944). A study of the mechanics of gastrulation. II. *J. Exp. Zool.* **95**, 171.

—— (1948). Concepts on the mechanism of embryonic induction and its relation to parthogenesis and malignancy. *Symp. Soc. Exp. Biol.* **2**, 17.

HOSKER, A. (1936). Studies on the epidermal structures of birds. *Phil. Trans. Roy. Soc.* B, **226**, 143.

HUBER, W. (1949). Analyse expérimentale des courbures de la tête chez le jeune embryon de poulet. *C.R. Ass. Anat.* **36**, 383.

HUGHES, A. F. W. (1935). Studies on the area vasculosa of the embryo chick. I. The first differentiation of the vitelline artery. *J. Anat.* **70**, 76.

—— (1937). Studies on the area vasculosa of the embryo chick. II. The influence of the circulation on the diameter of the vessels. *J. Anat.* **72**, 1.

HUNT, E. A. (1932). The differentiation of chick limb buds in chorio-allantoic grafts, with special reference to the muscles. *J. Exp. Zool.* **62**, 57.

HUNT, T. E. (1929). Hensen's node as an organiser in the formation of the chick embryo. *Anat. Rec.* **42**, 22.

—— (1931a). An experimental study of the independent differentiation of the isolated Hensen's node and its relation to the formation of axial and non-axial parts in the chick embryo. *J. Exp. Zool.* **59**, 395.

—— (1931b). Developmental potencies of explanted quadrants of Hensen's node. *Proc. Soc. Exp. Biol. Med.* **28**, 626.

—— (1932). Potencies of transverse levels of the chick blastoderm in the definitive-streak stage. *Anat. Rec.* **55**, 41.

—— (1937a). The development of gut and its derivatives from the mesectoderm and mesentoderm of early chick blastoderms. *Anat. Rec.* **68**, 349.

—— (1937b). The origin of entodermal cells from the primitive streak of the chick embryo. *Anat. Rec.* **68**, 449.

HYMAN, L. H. (1927). The metabolic gradients of vertebrate embryos. III. The chick. *Biol. Bull.* **52**, 1.

JACOBSON, W. (1938a). The early development of the avian embryo. I. Endoderm formation. *J. Morph.* **62**, 415.

JACOBSON, W. (1938b). The early development of the avian embryo. II. Mesoderm formation and the distribution of presumptive embryonic material. *J. Morph.* **62**, 445.

JACOBSON, W. & FELL, H. B. (1941). The developmental mechanics and potencies of the undifferentiated mesenchyme of the mandible. *Q. J. micr. Sci.* **82**, 563.

JONES, D. S. (1937). The origin of the sympathetic trunks in the chick embryo. *Anat. Rec.* **70**, 45.

—— (1939). Studies on the origin of sheath cells and sympathetic ganglia in the chick. *Anat. Rec.* **73**, 343.

—— (1941). Further studies on the origin of the sympathetic ganglia in the chick embryo. *Anat. Rec.* **79**, 7.

—— (1942). The origin of the vagi and the para-sympathetic ganglia cells of the viscera of the chick. *Anat. Rec.* **83**, 185.

JOY, E. A. (1939). Intra-coelomic grafts of the eye primordium of the chick. *Anat. Rec.* **74**, 461.

JUHN, M. & FRAPS, R. M. (1936). Developmental analysis in plumage. I. The individual feather: methods. *Physiol. Zool.* **9**, 293.

JUHN, M. & GUSTAVSON, R. G. (1930). The production of female genital subsidiary characters and plumage sex characters by injection of human placental hormone in fowls. *J. Exp. Zool.* **56**, 31.

KAUFMANN-WOLF, M. (1908). Embryologische und anatomische Beiträge zur Hyperdaktylie. *Morph. Jahrb.* **38**, 471.

KOCK, C. (1933). Capacita autodifferenziativa del cristallino dimonstrata negli impianti corio-allantoidei. *Monitore Zool. Ital.* **43**, 93.

KOLLER, P. C. (1929). Experimental studies on pigment formation. I. The development *in vitro* of the mesodermal pigment cells of the fowl. *Arch. Exp. Zellforsch.* **8**, 490.

KOPSCH, F. (1926a). Primitivstreifen und organbildende Keimbezirke beim Hühnchen untersucht mittels elektrolytischer Marken am vital gefärbten Keim. *Zeits. mikr.-anat. Forsch.* **8**, 512.

—— (1926b). Die Lage des Primitivstreifens im Hühnerei. *Zeits. mikr.-anat. Forsch.* **8**, 185.

—— (1934). Die Lage des Materials für Kopf, Primitivstreifen, und Gefasshof in der Keimscheibe des unbebrüteten Hühnereies und seine Entwicklung während der ersten beiden Tage der Bebrütung. *Zeits. mikr.-anat. Forsch.* **35**, 254.

KUME, M. (1935). The differentiating capacity of various regions of the heart rudiment of the chick as studied in chorio-allantoic grafts. *Physiol. Zool.* **8**, 73.

LALLEMAND, S. (1938). Realisation expérimentale à l'aide de la colchicine de poulet strophosomes. *C.R. Acad. Sci.* **207**, 1446.

—— (1939a). Action de la colchicine sur l'embryon de poulet à divers stades du développement. *C.R. Acad. Sci.* **208**, 1048.

—— (1939b). La strophosomie chez l'embryon de poulet. *Arch. Anat.* Strasbourg, **28**, 215.

LANDAUER, W. (1928). The morphology of intermediate rumplessness in the fowl. *J. Hered.* **19**, 453.

—— (1931). Untersuchungen über das Krüperhühn. II. Morphologie und Histologie des Skelets, insbesondere des Skelets der langen Extremitätenknochen. *Zeits. mikr.-anat. Forsch.* **25**, 115.

—— (1932). Studies on the Creeper fowl. III. The early development and lethal expression of homozygous Creeper embryos. *J. Genet.* **25**, 367.

—— (1933). Untersuchungen über das Krüperhuhn. IV. Die Missbildungen homozygoter Krüperembryonen auf späteren Entwicklungsstadien. *Zeits. mikr.-anat. Forsch.* **32**, 359.

—— (1934*a*). Studies on the Creeper fowl. VI. Skeletal growth of Creeper chickens. *Storrs Agric. Exp. Stat. Bull.* **193**.

—— (1934*b*). Studies on the Creeper fowl. VII. The expression of vitamin D deficiency (rickets) in Creeper chicks as compared with normal chicks. *Amer. J. Anat.* **55**, 229.

—— (1935*a*). Studies on the Creeper fowl. IX. Malformations occurring in the Creeper stock. *J. Genet.* **30**, 303.

—— (1935*b*). A lethal mutation in Dark Cornish fowl. *J. Genet.* **31**, 237.

—— (1936). Micromelia of chicken embryos and newly hatched chicks caused by a nutritional deficiency. *Anat. Rec.* **64**, 267.

—— (1937). Studies on the Creeper fowl. XI. Castration and length of bones of the appendicular skeleton in normal and Creeper fowl. *Anat. Rec.* **69**, 247.

—— (1939*a*). Studies on the Creeper fowl. XII. Size of body, organs and lung bones of late homozygous Creeper embryos. *Storrs Agric. Exp. Stat. Bull.* **232**.

—— (1939*b*). Studies on the lethal mutations of Cornish fowl. Growth in length of long bones and increase in weight of the body and of some organs. *Storrs Agric. Exp. Stat. Bull.* **233**.

—— (1940). Studies on the Creeper fowl. XIII. The effect of selenium and the asymmetry of selenium-induced malformations. *J. Exp. Zool.* **83**, 431.

—— (1941*a*). Teratological correlations and the mechanism of gene expression. *Proc. 7th int. Cong. Genet.* 181.

—— (1941*b*). A semi-lethal mutation in fowl affecting length of the upper beak and of the long bones. *Genetics*, **26**, 426.

—— (1941*c*). The hatchability of chicken eggs as influenced by environment and heredity. *Storrs Agric. Exp. Stat. Bull.* **236**.

—— (1944*a*). Length of survival of homozygous Creeper fowl embryos. *Sci.* **100**, 553.

—— (1944*b*). Transplantation as a tool of developmental genetics. *Amer. Nat.* **78**, 280.

—— (1945*a*). Recessive rumplessness of fowl with kyphoscoliosis and supernumerary ribs. *Genetics*, **30**, 403.

—— (1945*b*). Rumplessness of chicken embryos produced by the injection of insulin and other chemicals. *J. Exp. Zool.* **98**, 65.

—— (1946). Chondrodystrophy (Achondroplasia) and humoral agents. *Nature*, **157**, 838.

254 BIBLIOGRAPHY

LANDAUER, W. (1947a). Insulin-induced abnormalities of beak, extremities and eyes in chickens. *J. Exp. Zool.* **105**, 145.

—— (1947b). Potentiating effects of adrenal cortical extract on insulin-induced abnormalities of chick development. *Endocrinol.* **41**, 489.

—— (1947c). Insulin-induced rumplessness. V. The effect of insulin on the axial skeleton of chicks and adult fowl. *J. Exp. Zool.* **105**, 317.

—— (1948a). The phenotypic modification of hereditary polydactylism of fowl by selection and by insulin. *Genetics*, **33**, 133.

—— (1948b). The effect of nicotinamide and α-ketoglutamic acid on the teratogenic action of insulin. *J. Exp. Zool.* **109**, 283.

—— (1948c). Hereditary abnormalities and their chemically-induced phenocopies. *Growth Symp.* **12**, 171.

—— (1949). La problème de l'électivité dans les expériences de tératogenèse biochimique. *Arch. Anat. micr. Morph. exper.* **38**, 184.

LANDAUER, W. & BAUMANN, L. (1943). Rumplessness of chicken embryos produced by mechanical shaking of eggs prior to incubation. *J. Exp. Zool.* **93**, 51.

LANDAUER, W. & BLISS, C. I. (1946). Insulin-induced rumplessness of chickens. III. The relationship of dosage and of developmental stage at the time of injection to response. *J. Exp. Zool.* **102**, 1.

LANDAUER, W. & DUNN, L. C. (1930). Studies on the Creeper fowl. I. Genetics. *J. Genet.* **23**, 397.

LANDAUER, W. & LANG, E. H. (1946). Insulin induced rumplessness of chickens. II. Experiments with activated and reactivated insulin. *J. Exp. Zool.* **101**, 41.

LEHMANN, F. E. (1926). Entwicklungsstörungen an der Medullaranlage von Triton, erzeugt durch Unterlagerungsdefekte. *Arch. f. Entw. mech. Org.* **108**, 243.

—— (1938). Regionale Verschiedenheiten des Organisators von Triton. *Arch. f. Entw. mech. Org.* **138**, 106.

—— (1945). *Einführung in die physiologische Embryologie.* Basel: Birkhauser.

LERNER, A. B. & FITZPATRICK, T. B. (1950). The biochemistry of melanin formation. *Physiol. Rev.* **30**, 91.

LERNER, I. M. & GUNNS, C. A. (1937). Temperature and relative growth of chick embryo leg bones. *Growth*, **2**, 261.

LEVI-MONTALCINI, R. (1945). Corrélations dans le développement des différentes parties du système nerveux. *Arch. Biol.* **56**, 71.

—— (1947a). Richerche sulla correlazioni nello sviluppo del sistema nervoso. Regressione secondaria del ganglio ciliare dopo asportazione della vesicola mesencifalica in embrioni di pollo. *Atti Accad. Lincei*, Ser. 8A, **3**(1/2), 144.

—— (1947b). Richerche sperimentali sull' origine del simpatico toraco-lombare nell' embrione di pollo. *Atti Accad. Lincei*, Ser. 8A, **3**(1/2), 140.

—— (1949). The development of the acoustico-vestibular centers in the chick embryo in the absence of the different root fibers and of the descending fiber tracts. *J. Comp. Neuro.* **91**, 209.

LEVI-MONTALCINI, R. & AMPRINO, R. (1947). Recherches expérimentales sur l'origine du ganglion ciliaire dans l'embryon de poulet. *Arch. Biol.* **58**, 267.

LEVI-MONTALCINI, R. & LEVI, G. (1942). Les conséquences de la destruction d'une territoire d'innervation périphérique sur le développement des centres nerveux correspondants dans l'embryon de poulet. *Arch. Biol.* **53**, 537.

—— —— (1943). Recherches quantitatives sur la marche du processus de différenciation des neurones dans les ganglions spinaux de l'embryon de poulet. *Arch. Biol.* **54**, 189.

—— —— (1944). Correlazioni nello sviluppo tra varie parti del sistema nervoso. I. Consequenze della demolizione dell'abbozzo di un arto sui centri nervosi nell' embrione di pollo. *Comment. Pontif. Acad. Sci.* **8**, 527.

LILLIE, F. R. (1904). Experimental studies on the development of organs in the embryo of the fowl. II. The development of defective embryos and the power of regeneration. *Biol. Bull.* **7**, 33.

—— (1919). *The development of the chick.* 2nd. ed.[1] New York: Henry Holt.

—— (1929). Segregation and its role in life history. *Arch. f. Entw. mech. Org.* **118**, 499.

—— (1942). On the development of feathers. *Biol. Rev.* **17**, 247.

LILLIE, F. R. & JUHN, M. (1932). The physiology of development of feathers. I. Growth rate and pattern in the individual feather. *Physiol. Zool.* **5**, 124.

—— —— (1938). Physiology of development of the feather. II. General principles of development with special reference to the after-feather. *Physiol. Zool.* **11**, 434.

LILLIE, F. R. & WANG, H. (1941). Physiology of development of the feather. V. Experimental morphogenesis. *Physiol. Zool.* **14**, 103.

—— —— (1943). Physiology of development of the feather. VI. The production and analysis of feather chimaerae in the fowl. *Physiol. Zool.* **16**, 1.

—— —— (1944). Physiology of development of the feather. VII. An experimental study of induction. *Physiol. Zool.* **17**, 1.

LUTHER, W. (1934). Untersuchungen über die Umkehrbarkeit der Polarität zwischen Aussen- und Innenseite des Ektoderms von Amphibienkeimen. *Arch. f. Entw. mech. Org.* **131**, 532.

—— (1936). Potenzprüfungen an isolierten Teilstücken der Forellenkeimscheibe. *Arch. f. Entw. mech. Org.* **135**, 359.

—— (1937). Transplantations- und Defektversuche am Organisationszentrum der Forellenkeimscheibe *Arch. f. Entw. mech. Org.* **132**, 404.

LUTZ, H. (1948a). Sur la polyembryonie expérimentale resultant de fissurations en croix du blastoderme non incubé de cane. *C.R. Acad. Sci.* **226**, 841.

—— (1948b). Sur l'obtention expérimentale de la polyembryonie chez le canard. *C.R. Soc. Biol.* **143**, 384.

—— (1948c). Sur l'orientation des axes embryonnaires dans la polyembryonie expérimentale chez les oiseaux. *C.R. Soc. Biol.* **143**, 1016.

—— (1948d). Sur l'obtention expérimentale de monstres doubles chez les oiseaux. *C.R. Acad. Sci.* **227**, 87.

[1] A new edition is being prepared by H. L. Hamilton.

LUTZ, H. (1949). Sur la production expérimentale de la polyembryonie et de la monstrosité double chez les oiseaux. *Arch. Anat. micr. Morph. exper.* **38**, 79.

—— (1950a). L'influence du niveau de la section et du stade de l'incubation sur l'orientation des embryons doubles obtenus expérimentalement chez le canard. *C.R. Soc. Biol.* **144**, 1410.

—— (1950b). Sur l'orientation des embryons obtenus par fissuration perpendiculaire à l'axe presumé du blastoderme de l'œuf de cane. *C.R. Acad. Sci.* **231**, 379.

—— (1950c). Sur le conflit d'orientation des différentes parties de la moitié antérieure du blastoderme d'oiseaux. *C.R. Soc. Biol.* **144**, 1117.

LYONS, M. & INSKO, W. M. (1937). Chondrodystrophy in the chick embryo produced by manganese deficiency in the diet of the hen. *Kentucky Agric. Exp. Stat. Bull.* **371**.

MANGOLD, O. (1925). Die Bedeutung der Keimblätter in der Entwicklung. *Naturwiss.* **13**, 213.

—— (1931). Das Determinationsproblem. III. Das Wirbeltierauge in der Entwicklung und Regeneration. *Ergeb. Biol.* **7**, 196.

—— (1932). Autonome und komplementäre Induktionen bei Amphibien. *Naturwiss.* **20**, 371.

MANGOLD, O. & SPEMANN, H. (1927). Über Induktion von Medullarplatte durch Medullarplatte im jüngeren Keim. *Arch. f. Entw. mech. Org.* **111**, 341.

MARKERT, C. L. (1948). The effects of thyroxine and antithyroid compounds on the synthesis of pigment granules in chick melanoblasts cultivated *in vitro*. *Physiol. Zool.* **21**, 309.

MARTIN, F. R. (1943). Ejerce el prolan un efecto pernicioso sobre las primeras fases del desarrollo del embrion del pollo? *Rev. clin. Espanol.* **8**, 410.

MASON, H. S. (1948). A classification of melanins. *Spec. Publ., N.Y. Acad. Sci.* **4**, 399. (The Biology of Melanomas.)

MAXIMOV, A. (1925). Tissue culture of young mammalian embryos. *Contrib. Embryol. Carneg. Instn*, **16**, 47.

MAYER, E. (1942). Reversibility in the orientation of the chick embryo and the question of the *situs inversus viscerum*. *Anat. Rec.* **84**, 359.

McKEEHAN, M. S. (1950). Cytological aspects of embryonic induction in the chick embryo. *Anat. Rec.* **106**, 56.

McWHORTER, J. E. & WHIPPLE, H. O. (1912). The development of the blastoderm of the chick *in vitro*. *Anat. Rec.* **6**, 121.

MEDAWAR, P. B. (1947). Cellular inheritance and transformation. *Biol. Rev.* **22**, 360.

MEHRBACH, H. (1935). Beobachtungen an der Keimscheibe des Hühnchens vor dem Erscheinen des Primitivstreifens. *Zeits. Anat. Entw. ges.* **104**, 635.

MILFORD, J. J. (1941). Studies on homoplastic and heteroplastic transplantation of metanephric primordia to the coelom of the chick embryo. *Physiol. Zool.* **14**, 344.

MOOG, F. (1943a). Cytochrome oxidase in early chick embryos. *J. Cell. Comp. Physiol.* **22**, 223.

—— (1943b). The distribution of phosphatase in the spinal cord of chick embryos of one to eight days incubation. *Proc. Nat. Acad. Sci., Wash.*, **29**, 176.

Moog, F. (1944). Localisation of alkaline and acid phosphatases in the early embryogenesis of the chick. *Biol. Bull.* **86**, 51.

Morita, S. (1936). Die kunstliche Erzeugung von Einzelmissbildungen, von Zwillingen, Drillingen und Mehrlingen im Hühnerei. *Anat. Anz.* **82**, 81.

—— (1937). Die kunstliche Erzeugung von Mehrfachbildungen im Hühnerei. *Anat. Anz.* **84**, 81.

Moseley, H. R. (1947). Insulin-induced rumplessness of chickens. IV. Early embryology. *J. Exp. Zool.* **105**, 279.

Murray, P. D. F. (1926). An experimental study of the development of the limbs of the chick. *Linn. Soc. New South Wales*, **51**, 179.

—— (1928a). Chorio-allantoic grafts of fragments of the two-day chick, with special reference to the development of the limbs, intestine and skin. *Austral. J. Exp. Biol. Med. Sci.* **5**, 237.

—— (1928b). The origin of the dermis. *Nature*, **122**, 609.

—— (1932). The development *in vitro* of the blood of the early chick embryo. *Proc. Roy. Soc. B*, **111**, 497.

—— (1933). The cultivation in saline and other media of the haematopoietic region of the early chick embryo. *Arch. Exp. Zellf.* **14**, 574.

Murray, P. D. F. & Huxley, J. S. (1925). Self-differentiation in the grafted limb-bud of the chick. *J. Anat.* **59**, 379.

Murray, P. D. F. & Selby, D. (1930a). Chorio-allantoic grafts of entire and fragmented blastoderms of the chick. *J. Exp. Biol.* **7**, 404.

—— —— (1930b). Intrinsic and extrinsic factors in the primary development of the skeleton. *Arch. f. Entw. mech. Org.* **122**, 629.

Needham, J. (1931). *Chemical embryology.* 3 vols. Cambridge Univ. Press.

—— (1933). On the dissociability of the fundamental processes in ontogenesis. *Biol. Rev.* **8**, 180.

—— (1942). *Biochemistry and morphogenesis.* Cambridge Univ. Press.

Needham, J. & Nowinski, W. W. (1937). Intermediary carbohydrate metabolism in embryonic life. I. General aspects of anaerobic glycolysis. *Biochem. J.* **31**, 1165.

Needham, J. Waddington, C. H. & Needham, D. M. (1934). Physico-chemical experiments on the amphibian organiser. *Proc. Roy. Soc. B*, **114**, 393.

Newman, H. H. (1940). Twin and triplet chick embryos. *J. Hered.* **31**, 371.

Nickerson, M. (1944). An experimental analysis of barred pattern in feathers. *J. Exp. Zool.* **95**, 361.

Nieuwkoop, P. D. & Florschütz, P. A. (1950). Quelques caractères spéciaux de la gastrulation et de la neurulation de l'oeuf de *Xenopus laevis.* Pt. I. *Arch. Biol.* **61**, 113.

Niu, M. C. (1947). The axial organisation of the neural crest. *J. Exp. Zool.* **105**, 79.

Niven, J. S. F. (1933). The development *in vivo* and *in vitro* of the avian patella. *Arch. f. Entw. mech. Org.* **128**, 480.

Northrop, F. S. C. & Burr, H. S. (1937). Fundamental findings concerning the electrodynamic theory of life and an analysis of their physical meaning. *Growth*, **1**, 78.

OAKLEY, C. L. (1938). Chorio-allantoic grafts of liver. *J. Path. & Bact.* **46**, 109.

OLIVO, O. M. (1928). Précoce détermination de l'ébauche du coeur dans l'embryon de poulet et sa différenciation histologique et physiologique *in vitro*. *C.R. Ass. Anat.* **23**, 357.

OLSEN, M. W. (1942). Maturation, fertilisation and early cleavage in the hen's egg. *J. Morph.* **70**, 513.

OPPENHEIMER, J. M. (1934). Experiments on early developing stages of Fundulus. *Proc. Nat. Acad. Sci.*, Wash., **20**, 536.

—— (1936a). Processes of localisation in developing Fundulus. *J. Exp. Zool.* **73**, 405.

—— (1936b). The development of isolated blastoderms of *Fundulus hetero-clitus*. *J. Exp. Zool.* **72**, 247.

ORTS LIORCA, F. (1943). Erzeugung von Asymmetrien im Hühnerei durch Einwirkung von Testoviron. *Arch. f. Entw. mech. Org.* **142**, 619.

—— (1944). Acción perturbadora del testovirón y sales de litio sobre los primeros estudios del desarollo en el embrión de pollo. *Arch. Espanol Morphol.* **10**, 131.

—— (1948). Evocación de placa y canal neural en el blastodermo de pollo, por medio de testículo humano muerto por ebullición y desecado. *Rev. Espan. Oto-Neuro-Optalmol. y Neurocirurgia*, **40**, 1.

PAFF, G. H. (1939). The action of colchicine upon the 48-hour chick embryo. *Amer. J. Anat.* **64**, 331.

PASTEELS, J. (1935a). Les mouvements morphogénétiques de la gastrulation chez les Tortues. *Bull. Classe Sci. Acad. Roy. Belg.* **21**, 88.

—— (1935b). Sur les mouvements morphogénétiques suscitant l'apparition de la ligne primitive chez les oiseaux. *C.R. Soc. biol. Belg.* **120**, 1362.

—— (1936a). Analyse des mouvements morphogénétiques de gastrulation chez les oiseaux. *Bull. Classe Sci. Acad. Roy. Belg.* **22**, 737.

—— (1936b). Études sur la gastrulation des Vertébrés méroblastiques. I. Téléostéens. *Arch. Biol.* **47**, 205.

—— (1936c). Centre organisateur et glycogénolyse. *Arch. Anat. micr.* **32**, 303.

—— (1937). Études sur la gastrulation des Vertébrés méroblastiques II. Reptiles. III. Oiseaux. IV. Conclusions générales. *Arch. Biol.* **48**, 105.

—— (1939). La formation de la queue chez les Vertébrés. *Ann. Soc. Roy. Zool. Belg.* **70**, 33.

—— (1938–41). Recherches sur les facteurs initiaux de la morphogénése chez les amphibiens anoures. I–VII. *Arch. Biol.* **49, 50, 51, 52**.

—— (1940). Un aperçu comparatif de la gastrulation chez les Chordes. *Biol. Rev.* **15**, 59.

—— (1942a). Sur l'existence éventuelle d'une croissance au cours de la gastrulation des Vertébrés. *Acta biol. belg.* **1**, 126.

—— (1942b). Le bourgeon caudal joue-t-il un rôle dans la croissance du jeune embryon des Vertébrés? *Acta biol. belg.* **1**, 130.

—— (1943). Prolifération et croissance dans la gastrulation et la formation de la queue des Vertébrés. *Arch. Biol.* **54**, 1.

PASTEELS, J. (1945). On the formation of the primary entoderm of the duck (*Anas domestica*) and on the significance of the bilaminar embryo in birds. *Anat. Rec.* **93**, 5.

—— (1947). Sur l'apparition d'organes variés dans l'ectoblaste, à la suite de la centrifugation de la blastula et de la gastrula chez les Amphibiens. *Experientia*, **3**(i), 30.

—— (1948). Production d'embryos surnuméraires et de tératones chez les Amphibiens par la centrifugation. *C.R. Soc. Biol.* **142**, 1320.

—— (1949). Observations sur la localisation de la plaque préchordale et de l'entoblaste présomptifs au cours de la gastrulation chez *Xenopus laevis*. *Arch. Biol.* **60**, 235.

PATTEN, B. M. (1949). Initiation and early changes in the character of the heart beat in vertebrate embryo. *Physiol. Rev.* **29**, 31.

PATTERSON, J. F. (1909). Gastrulation in the pigeon's egg. *J. Morph.* **20**, 65.

PEEBLES, F. (1898). Some experiments on the primitive streak of the chick. *Arch. f. Entw. mech. Org.* **7**, 405.

—— (1904). The location of the chick embryo upon the blastoderm. *J. Exp. Zool.* **1**, 369.

—— (1911). On the interchange of the limbs of the chick by transplantation. *Biol. Bull.* **20**, 14.

PETER, K. (1934). Indirekte (primäre) und direkte (sekundäre) Körperentwicklung bei der Eidechse. *Zeits. mikr.-anat. Forsch.* **36**, 378.

—— (1938a). Untersuchungen über die Entwicklung des Dotterentoderms. I. Die Entwicklung des Entoderms beim Hühnchen. *Zeits. mikr.-anat. Forsch.* **43**, 362.

—— (1938b). Untersuchungen über die Entwicklung des Dotterentoderms. II. Die Entwicklung des Entoderms bei der Taube. *Zeits. mikr.-anat. Forsch.* **43**, 416.

—— (1938c). Untersuchungen über die Entwicklung des Dotterentoderms. III. Die Entwicklung des Entoderms bei Reptilien. *Zeits. mikr.-anat. Forsch.* **44**, 498.

—— (1939). Untersuchungen über die Entwicklung des Dotterentoderms. IV. Das Schicksal des Dotterentoderms beim Hühnchen. *Zeits. mikr.-anat. Forsch.* **46**, 627.

PHILLIPS, F. S. (1941). The oxygen consumption of the early chick embryo at various stages of development. *J. Exp. Zool.* **86**, 257.

—— (1942). Comparison of the respiratory rates of different regions of the chick blastoderm during early stages of development. *J. Exp. Zool.* **90**, 83.

PIATT, J. (1948). Form and causality in neurogenesis. *Biol. Rev.* **23**, 1.

POMERAT, C. M. (1949). Morphogenetic effects of spleen antigen and antibody administrations to chick embryos. *Exper. Cell. Res.* Suppl. **1**, 578.

RAVEN, C. P. (1933). Zur Entwicklung der Ganglienleiste. II. Über das Differenzierungsvermögen des Kopfganglienleistenmaterials von Urodelen. *Arch. f. Ent. mech. Org.* **129**, 179.

—— (1948). On the concepts of experimental embryology. *Fol. Biotheoretica*, B, **3**, 1.

RAWLES, M. E. (1936). A study in the localization of organ-forming areas in the chick blastoderm of the head-process stage. *J. Exp. Zool.* **72**, 271.

—— (1939). The production of robin pigment in White Leghorn feathers by grafts of embryonic robin tissue. *J. Genet.* **38**, 517.

—— (1940). The pigment-forming potency of early chick blastoderms. *Proc. Nat. Acad. Sci., Wash.*, **26**, 86.

—— (1943). The heart forming areas of the early chick blastoderm. *Physiol. Zool.* **16**, 22.

—— (1944). The migration of melanoblasts after hatching into pigment-free skin grafts of the common fowl. *Physiol. Zool.* **17**, 167.

—— (1947). Some observations on the developmental properties of the presumptive hind limb area of the chick. *Anat. Rec.* **99**, 648.

—— (1948). Origin of melanophores and their role in development of color patterns in vertebrates. *Physiol. Rev.* **28**, 383.

RAWLES, M. E. & STRAUS, W. L. (1948). An experimental analysis of the development of the trunk musculature and ribs in the chick. *Anat. Rec.* **100**, 755.

REMOTTI, E. (1927). Ricerche fisio-morfologiche sul sacco vitellino del pollo. *Ric. Morphol.* **7**.

—— (1931). Sulla riduzione del sacco vitellino negli uccelli. *Ric. Morphol.* **11**.

—— (1933*a*) Sulla sopravivenza dell sacco vitellino degli uccelli alla morte dell' embrione. *Bol. Phis. Lab. Zool., Anat., Comp. Genova*, **13**.

—— (1933*b*). Su alcuni fenomeni di disarmonia di sviluppo embrionale. *Monit. Zool. Ital.* **43**, 218.

—— (1933*c*). Il sacco vitellino degli uccelli in condizione di persistenza sperimentali. *Ric. Morphol.* **12**.

REVERBERI, G. (1929*a*) Risulti di esperimenti sullo sviluppo dell' occhio nell' embrione di pollo. *R.C. Accad. Lincei, Ser.* VI*a*, **10**, 115.

—— (1929*b*). Risulti di esperimenti di asportazione parziale e totale della vesicola ottica nell' embrione di pollo. *Boll. Inst. Zool. Univ. di Roma*, **7**, 1.

—— (1930). Sulla formazione della lente dal margine del calice ottico nell' embrione di pollo. *Arch. Zool. Ital.* **15**, 337.

RHINES, R. (1944). Formation of commissures in surgically altered brains of chick embryos. *Anat. Rec.* **88**, 454.

RHINES, R. & WINDLE, W. F. (1944). An experimental study of factors influencing the course of nerve fibres in the embryonic nervous system. *Anat. Rec.* **90**, 267.

RHOADS, C. P. (1949). Neoplastic abnormal growth, in *The chemistry and physiology of growth*. Princeton Univ. Press.

RIS, H. (1941). An experimental study on the origin of melanophores in birds. *Physiol. Zool.* **14**, 48.

ROMANOFF, A. L. & BAUERNFEIND, J. C. (1942). Influence of riboflavin deficiency in eggs on embryonic development (*Gallus domesticus*). *Anat. Rec.* **82**, 11.

ROMANOFF, A. L. & ROMANOFF, A. J. (1949). *The avian egg*. New York: Wiley & Sons; London: Chapman & Hall.

Rose, S. M. (1939). Embryonic induction in the Ascidia. *Biol. Bull.* **77**, 216.

Rous, P. & Murphy, J. B. (1911). Tumor implantations in the developing embryo. *J. Amer. Med. Assoc.* **56**, 741.

Rudnick, D. (1932). Thyroid forming potencies of the early chick blastoderm. *J. Exp. Zool.* **62**, 287.

—— (1933). Developmental capacities of the chick lung in chorio-allantoic grafts. *J. Exp. Zool.* **66**, 125.

—— (1935). Regional restriction of potencies in the chick during embryogenesis. *J. Exp. Zool.* **71**, 83.

—— (1938a). Contribution to the problem of neurogenic potency in postnodal isolates from chick blastoderm. *J. Exp. Zool.* **78**, 369.

—— (1938b). Differentiation in culture of pieces of the early chick blastoderm. I. The definitive primitive streak and head-process stages. *Anat. Rec.* **70**, 351.

—— (1938c). Differentiation in culture of pieces of the early chick blastoderm. II. Short primitive streak stages. *J. Exp. Zool.* **89**, 399.

—— (1944). Early history and mechanics of the chick blastoderm. A review. *Quart. Rev. Biol.* **19**, 187.

—— (1945a). Differentiation of prospective limb material from Creeper chick embryos in coelomic grafts. *J. Exp. Zool.* **100**, 1.

—— (1945b). Limb-forming potencies of the chick blastoderm: including notes on associated trunk structures. *Trans. Conn. Acad. Arts and Sci.* **36**, 353.

—— (1948). Prospective areas and differentiation potencies in the chick blastoderm. *Ann. N.Y. Acad. Sci.* **49**, 761.

Rudnick, D. & Hamburger, V. (1940). On the identification of segregated phenotypes in progeny from Creeper fowl matings. *Genetics*, **25**, 215.

Rudnick, D. & Rawles, M. E. (1937). Differentiation of the gut in chorio-allantoic grafts from chick blastoderms. *Physiol. Zool.* **10**, 381.

Rulon, O. (1935). Differential reduction of Janus Green during development of the chick. *Protoplasma*, **24**, 346.

Ruud, G. & Spemann, H. (1922). Die Entwicklung isolierter dorsaler und lateraler Gastrulahälften von Triton. *Arch. f. Entw. mech. Org.* **52**, 95.

Sabin, F. (1919). Studies in the origin of blood vessels and of red blood-corpuscles as seen in the living blastoderm of chicks during the second day of incubation. *Carn. Inst. Wash. Publ.* **272**.

Saunders, J. W., Jr. (1948a). The proximo-distal sequence of origin of the chick wing and the role of the ectoderm. *J. Exp. Zool.* **108**, 363.

—— (1948b). Do the somites contribute to the formation of the chick wing? *Anat. Rec.* **100**, 756.

—— (1949a). Analysis of the role of the apical ridge of ectoderm in the development of the limb-bud in the chick. *Anat. Rec.* **105**, 567.

—— (1949b). Can prospective pigment cells of the chick embryo be identified by means of vital dyes during the course of their migration into the skin and feather germs? *Anat. Rec.* **105**, 596.

Schechtman, A. M. (1947). Antigens of early developmental stages of the chick. *J. Exp. Zool.* **105**, 329.

SCHMIDT, G. (1937). On the growth stimulating effect of egg-white and its importance for embryonic development. *Enzymologia*, **4**, 40.

SEEVERS, C. H. (1932). Potencies of the end-bud and other caudal levels of the early chick embryo, with special reference to the origin of the metanephros. *Anat. Rec.* **54**, 217.

SELBY, D. & MURRAY, P. D. F. (1928). Grafts of longitudinal halves of limb-buds of the four-day chick. *Austral. J. Exp. Biol. Med. Sci.* **5**, 181.

SIMMLER, G. M. (1949). The effects of wing bud extirpation on the brachial sympathetic ganglia of the chick embryo. *J. Exp. Zool.* **110**, 247.

SORIN, A. N. (1928). Über mitogenetische Induktion in den frühen Entwicklungstadien des Hühnerembryos. *Arch. f. Entw. mech. Org.* **113**, 724.

SPEMANN, H. (1938). *Embryonic Development and Induction.* Yale Univ. Press.

SPEMANN, H. & GEINITZ, B. (1927). Über Weckung organisatorischer Fähigkeiten durch Verpflanzung in organisatorische Umgebung. *Arch. f. Entw. mech. Org.* **109**, 129.

SPIEGELMAN, S. (1945). Physiological competition as a regulatory mechanism in morphogenesis. *Q. Rev. Biol.* **20**, 121.

SPRATT, N. T. (1940). An *in vitro* analysis of the organization of the eye-forming area in the early chick blastoderm. *J. Exp. Zool.* **85**, 171.

—— (1942). Location of organ-specific regions and their relationship to the development of the primitive streak in the early chick blastoderm. *J. Exp. Zool.* **89**, 69.

—— (1946). Formation of the primitive streak in the explanted chick blastoderm marked with carbon particles. *J. Exp. Zool.* **103**, 259.

—— (1947a). Localisation of prospective chorda and somite mesoderm during regression of the primitive streak in the chick blastoderm. *Anat. Rec.* **99**, 653.

—— (1947b). Localisation of the prospective neural plate in the primitive streak blastoderm of the chick. *Anat. Rec.* **99**, 654.

—— (1947c). Regression and shortening of the primitive streak in the explanted chick blastoderm. *J. Exp. Zool.* **104**, 69.

—— (1947d). A simple method for explanting and cultivating early chick embryos *in vitro*. *Science*, **106**, 452.

—— (1947e). Development *in vitro* of the early chick blastoderm explanted on yolk and albumen extract saline-agar substrata. *J. Exp. Zool.* **106**, 345.

—— (1948a). Development of the early chick blastoderm on synthetic media. *J. Exp. Zool.* **107**, 39.

—— (1948b). Nutritional requirements of the early chick embryo. I. The utilisation of carbohydrate substances. *Anat. Rec.* **101**, 65.

SPURLING, R. G. (1923). The effect of extirpation of the posterior limb bud on the development of the limb and pelvic girdles in chick embryos. *Anat. Rec.* **26**, 41.

STCHERBATOV, I. I. (1938). Transplantation of auditory vesicle of chick embryo into chorio-allantois. *Bull. biol. et Med. exper. U.R.S.S.* **6**, 511.

STEIN, K. F. (1929). Early embryonic differentiation of the chick hypophysis as shown in chorio-allantoic grafts. *Anat. Rec.* **43**, 221.

STEIN, K. F. (1933). The location and differentiation of the presumptive ectoderm of the forebrain and hypophysis as shown by chorio-allantoic grafts. *Physiol. Zool.* **6**, 205.

STÉPHAN, F. (1949*a*). Sur la ligature des arcs aortiques chez l'embryon de Poulet. *C.R. Soc. Biol.* **143**, 291.

—— (1949*b*). Les suppléances obtenues expérimentalement dans le système des arcs aortiques de l'embryon d'oiseau. *C.R. Ass. Anat.* **36**, 647.

STOCKENBERG, W. (1937). Die Orte besonderer Vitalfärbbarkeit des Hühnerembryos und ihre Bedeutung für die Formbildung. *Arch. f. Entw. mech. Org.* **135**, 408.

STRANGEWAYS, T. S. P. & FELL, H. B. (1926). Experimental studies on the differentiation of embryonic tissues growing *in vivo* and *in vitro*. II. The development of the isolated early embryonic eye of the fowl when cultivated *in vitro*. *Proc. Roy. Soc.* B, **100**, 273.

—— —— (1929). A study of the direct and indirect action of X-rays upon the tissues of the embryonic fowl. *Proc. Roy. Soc.* B, **102**, 9.

STREET, S. F. (1937). The differentiation of the nasal area of the chick embryo in grafts. *J. Exp. Zool.* **77**, 49.

STUDITSKY, A. N. (1944). Agents responsible for achondroplasia foetalis. *C.R. (Doklady) Acad. Sci. U.R.S.S.* **43**, 391.

—— (1946). Endocrine correlation in the embryonal development of the vertebrates. *Nature*, **157**, 427.

STURKIE, P. D. (1943). Suppression of polydactyly in the domestic fowl by low temperatures. *J. Exp. Zool.* **93**, 325.

—— (1946). The production of twins in *Gallus domesticus*. *J. Exp. Zool.* **101**, 51.

SZEPSENWOL, J. (1933). Recherches sur les centres organisateurs des vésicules auditives chez des embryons de poulets omphalocéphales obtenus expérimentalement. *Arch. Anat. micr.* **29**, 5.

—— (1934). Les conditions embryologiques qui mènent à la formation d'un coeur unique ou double. *Arch. Anat., Strasbourg*, **17**, 307.

—— (1940*a*). Diferenciación de las células motoras de la médula en cultivos *in vitro*. *Rev. Soc. Argent. Biol.* **16**, 352.

—— (1940*b*). El trajecto de las fibras nerviosas radiculares y fasciculares en cultivos *in vitro*. *Rev. Soc. Argent. Biol.* **16**, 589.

—— (1940*c*). Influencia de los somitos sobre la diferenciación y el trajecto de los nervios *in vitro*. *Rev. Soc. Argent. Biol.* **16**, 608.

SZEPSENWOL, J. & GOLDSTEIN, S. (1938). Différenciation *in vitro* des cellules nerveuses jeunes. *Arch. exp. Zellforsch.* **21**, 155.

TAYLOR, K. M. & SCHECHTMAN, A. M. (1949). *In vitro* development of the early chick embryo in the absence of small organic molecules. *J. Exp. Zool.* **111**, 227.

TAYLOR, L. W. & GUNNS, C. A. (1947). Diplopodia; a lethal form of polydactyly in chickens. *J. Hered.* **38**, 67.

TAZELAAR, M. A. (1928). The effect of a temperature gradient on the early development of the chick. *Q. J. micr. Sci.* **72**, 419.

TRINKAUS, J. P. (1948). Factors concerned in the response of melanoblasts to oestrogen in the Brown Leghorn fowl. *J. Exp. Zool.* **109**, 135.

TSANG, Y. C. (1939). Ventral horn cells and polydactyly in mice. *J. Comp. Neurol.* **70**, 1.

TUNG, T. C., CHANG, C. Y. & TUNG, Y. F. Y. (1945). Experiments on the developmental potencies of blastoderms and fragments of teleostean eggs separated latitudinally. *Proc. Zool. Soc. Lond.* **115**, 175.

TWIESSELMANN, F. (1935). Production d'embryons doubles, par une lésion electrolytique du blastodisque non incubé de poulet. *C.R. Soc. Biol.* **119**, 1169.

—— (1938). Expériences de scission précoce de l'aire embryogène chez le Poulet. *Arch. Biol.* **49**, 285.

TWITTY, V. C. (1936). Correlated genetic and embryological experiments on Triturus. *J. Exp. Zool.* **74**, 239.

—— (1949). Developmental analysis of amphibian pigmentation. *Growth*, Suppl. to Vol. **13**, 133.

UBISCH, L. v. (1939). Über die Entwicklung von Ascidienlarven nach frühzeitiger Entfernung der einzelnen organbildenden Keimbezirke. *Arch. f. Entw. mech. Org.* **139**, 438.

UMANSKI, E. (1928). Einige Bemerkungen über Elektrizitätseinwirkung auf die Entwicklung des Embryos von *Gallus domesticus*. *Zool. Anz.* **79**, 27.

—— (1931). Das Organisationszentrum der Primitiventwicklung von *Gallus domesticus*. *Zool. Anz.* **96**, 299.

VALENTIN, G. (1851). Ein Beitrag zur Entwicklungsgeschichte der Doppelmissbildungen. *Vierordt's Arch. f. phys. Heilk.* **1**.

VANDEBROEK, G. (1936). Les mouvements morphogénétiques au cours de la gastrulation chez *Scyllium Canicula* (Cav.). *Arch. Biol.* **47**, 499.

—— (1939). Cf. REVERBERI, G. (1948). Nouveaux résultats et nouvelles vues sur la germe des Ascidies. *Fol. Biotheoret.* **3**, 59.

VISINTINI, F. & LEVI-MONTALCINI, R. (1939). Relazione tra differenziazione strutturale e funzionale dei centri e delle vie nervose nell' embrione di pollo. *Arch. Suisses Neur. et Psych.* **44**, 119.

VOLKER, O. (1944). Die stofflichen Grundlagen der Pigmentierung der Vögel. *Biol. Zentralbl.* **84**, 184.

WADDINGTON, C. H. (1930). Developmental mechanics of chick and duck embryos. *Nature*, **125**, 924.

—— (1932). Experiments on the development of chick and duck embryos, cultivated *in vitro*. *Phil. Trans. Roy. Soc.* B, **221**, 179.

—— (1933a). Induction by the primitive streak and its derivatives in the chick. *J. Exp. Biol.* **10**, 38.

—— (1933b). Induction by coagulated organisers in the chick embryo. *Nature*, **131**, 275.

—— (1933c). Induction by the endoderm in birds. *Arch. f. Entw. mech. Org.* **128**, 502.

—— (1934a). Experiments on embryonic induction. I. The competence of the extra-embryonic ectoderm in the chick. *J. Exp. Biol.* **11**, 211.

WADDINGTON, C.H. (1934*b*). Experiments on embryonic induction. II. Experiments on coagulated organisers in the chick. *J. Exp. Biol.* **11**, 218.

—— (1934*c*). Experiments on embryonic induction. III. A note on inductions by the chick primitive streak transplanted to the rabbit embryo. *J. Exp. Biol.* **11**, 224.

—— (1935). The development of isolated parts of the chick blastoderm. *J. Exp. Zool.* **71**, 273.

—— (1937*a*). The dependence of head curvature on the development of the heart in the chick embryo. *J. Exp. Biol.* **14**, 229.

—— (1937*b*). The determination of the auditory placode in the chick. *J. Exp. Biol.* **14**, 232.

—— (1937*c*). Experiments on determination in the rabbit embryo. *Arch. Biol.* **48**, 273.

—— (1938*a*). The morphogenetic function of a vestigial organ in the chick. *J. Exp. Biol.* **15**, 371.

—— (1938*b*). Regulation of amphibian gastrulae with added ectoderm. *J. Exp. Biol.* **15**, 377.

—— (1939). The organisation centre of the chick embryo. *Current Science,* Special Number, August, p. 39.

—— (1940). *Organisers and genes.* Cambridge Univ. Press.

—— (1941). Twinning in chick embryos. *J. Hered.* **32**, 268.

—— (1942). Observations on the forces of morphogenesis in the amphibian embryo. *J. Exp. Biol.* **19**, 284.

—— (1948). The genetic control of development. *Symp. Soc. Exp. Biol.* **2**, 145.

—— (1949). The concept of organisation. *Proc. Xth Int. Cong. Philos.* p. 652.

—— (1950). Processes of induction in the early development of the chick. *Année Biol.* **26**, 711.

WADDINGTON, C. H. & COHEN, A. (1936). Experiments on the development of the head of the chick embryo. *J. Exp. Biol.* **13**, 219.

WADDINGTON, C. H., NEEDHAM, J. & BRACHET, J. (1936). Studies on the nature of the amphibian organisation centre. *Proc. Roy. Soc. B,* **120**, 173.

WADDINGTON, C. H. & SCHMIDT, G. A. (1933). Induction by heteroplastic grafts of the primitive streak in birds. *Arch. f. Entw. mech. Org.* **128**, 522.

WADDINGTON, C. H. & TAYLOR, J. (1937). Conversion of presumptive ectoderm to mesoderm in the chick. *J. Exp. Biol.* **14**, 335.

WADDINGTON, C. H. & YAO, T. (1950). Studies on regional specificity in the amphibian organisation centre. *J. Exp. Biol.* **27**, 126.

WANG, H. (1943). The morphogenetic functions of the epidermal and dermal components of the papilla in feather regeneration. *Physiol. Zool.* **16**, 325.

—— (1945). Experimental alteration of growth rate in chimaeric feathers of breast-saddle origin in the Brown Leghorn capon. *Physiol. Zool.* **18**, 335.

WARREN, A. E. (1934). Experimental studies on the development of the wing in the embryo of *Gallus domesticus. Amer. J. Anat.* **54**, 449.

WARREN, D. C. (1941). A new type of polydactyly in the fowl. *J. Hered.* **32**, 3.

—— (1944). Inheritance of polydactyly in the fowl. *Genetics,* **29**, 217.

WATERMAN, A. J. (1936). Experiments on young chick embryos cultured *in vitro*. *Proc. Nat. Acad. Sci., Wash.*, **22**, 1.

WATERMAN, A. J. & EVANS, H. J. (1940). Morphogenesis of the avian ear rudiment in chorio-allantoic grafts. *J. Exp: Zool.* **84**, 53.

WATTERSON, R. L. (1942). The morphogenesis of down feathers with special reference to the developmental history of melanophores. *Physiol. Zool.* **15**, 234.

WEBER, A. (1941). Action inhibitrice de l'ébauche oculaire sur l'évolution de la chorde dorsale chez les embryons de poulet. *Rev. suisse Zool.* **48**, 339.

WEEL, P. B. (1948). Histophysiology of the limb bud of the fowl during its early development. *J. Anat.* **82**, 49.

WEISS, P. (1930). *Entwicklungsphysiologie der Tiere*. Dresden and Leipzig.

—— (1939). *Principles of Development*. New York: Henry Holt.

—— (1941). Nerve patterns; the mechanics of nerve growth. *Growth*, **3**, 163.

—— (1945). Experiments on cell and axon orientation *in vitro*: the role of colloidal exudates in tissue organisation. *J. Exp. Zool.* **100**, 353.

—— (1947). The problem of specificity in growth and development. *Yale J. Biol. Med.* **19**, 235.

—— (1950a). *Genetic neurology*. (Ed.) Univ. Chicago Press.

—— (1950b). Perspectives in the field of morphogenesis. *Q. Rev. Biol.* **25**, 177.

WETZEL, R. (1924). Über den Primitivknoten des Hühnchens. *Verh. d. phys. med. Ges. Würzburg.* **49**, 227.

—— (1926). Wachstumszentrum und Kopfproblem in der ersten Entwicklung des Hühns. *Anat. Anz.* **61**, 87.

—— (1929a). Untersuchungen am Hühnchen. Die Entwicklung des Keims während der ersten beiden Bruttage. *Arch. f. Entw. mech. Org.* **119**, 188.

—— (1929b). Neue Experimente zur Frühentwicklung des Hühnes. *Anat. Anz.* **67**, 76.

—— (1931). Urmund und Primitivstreifen. *Erg. Anat. Entwges.* **29**, 1.

—— (1936). Primitivstreifen und Urkörper nach Störungsversuchen am 1–2 Tage bebrüteten Hühnchen. *Arch. f. Entw. mech. Org.* **134**, 357.

WHITE, P. R. (1946). Cultivation of animal tissues *in vitro* in nutrients of precisely known constitution. *Growth*, **10**, 231.

WILLIAMS, R. G. (1931). A study of the growth of a portion of the spinal cord following its early isolation from the central nervous system in the chick embryo. *J. Comp. Neurol.* **52**, 255.

WILLIER, B. H. (1924). The endocrine glands and the development of the chick. *Amer. J. Anat.* **33**, 67.

—— (1930). A study of the origin and differentiation of the suprarenal gland in the chick embryo by chorio-allantoic grafting. *Physiol. Zool.* **3**, 201.

—— (1942). Hormonal control of embryonic differentiation in birds. *Cold Spring Harbor Symp.* **10**, 135.

—— (1948). Hormonal regulation of feather pigmentation in the fowl. *Spec. Publ. N.Y. Acad. Sci.* **4**, 321. (The Biology of Melanomas).

WILLIER, B. H. & RAWLES, M. E. (1931a). Developmental relations of the heart and liver in chorio-allantoic grafts of whole chick blastoderm. *Anat. Rec.* **48**, 277.

—— —— (1931b). The relation of Hensen's node to the differentiating capacity of whole chick blastoderms as studied in chorio-allantoic grafts. *J. Exp. Zool.* **59**, 429.

—— —— (1935). Organ-forming areas in the early chick blastoderm. *Proc. Soc. Exp. Biol. Med.* **32**, 1293.

—— —— (1938). Feather characterization as studied in host-graft combinations between chick embryos of different breeds. *Proc. Nat. Acad. Sci., Wash.*, **24**, 446.

—— —— (1940). The control of feather color pattern by melanophores grafted from one embryo to another of a different breed of fowl. *Physiol. Zool.* **13**, 177.

—— —— (1944). Genotypic control of feather color pattern as demonstrated by the effects of a sex-linked gene upon melanophores. *Genetics*, **29**, 309.

WILLIER, B. H., RAWLES, M. E. & HADORN, E. (1937). Skin transplants between embryos of different breeds of fowl. *Proc. Nat. Acad. Sci., Wash.*, **23**, 10.

WOERDEMANN, M. W. (1933). Über den Glycogenstoffwechsel des Organisationszentrum in der Amphibiengastrula. *Proc. Acad. Sci. Amst.* **36**, 189.

WOLFF, E. (1932). Sur l'indépendance du développement de l'amnion et de l'embryon chez le poulet. *C.R. Soc. Biol.* **111**, 740.

—— (1933a). Recherches sur la structure d'omphalocéphales obtenus expérimentalement. *Arch. Anat., Strasbourg*, **16**, 155.

—— (1933b). Les stades précoces de l'omphalocéphalie. *C.R. Soc. Biol.* **112**, 805.

—— (1933c). La topographie des ébauches présomptives du foie d'après l'étude des Poulets omphalocéphales. *C.R. Acad. Sci.* **196**, 431.

—— (1934a). Les conséquences de la lésion de la région du nœud de Hensen sur le développement du poulet. *C.R. Soc. Biol.* **118**, 77.

—— (1934b). Production expérimentale de la symélie chez le poulet. *C. R. Soc. Biol.* **116**, 780.

—— (1934c). Recherches expérimentales sur la cyclopie. *Arch. Anat., Strasbourg*, **18**, 145.

—— (1934d). Recherches expérimentales sur l'otocéphalie et les malformations fondamentales de la face. *Arch. Anat., Strasbourg*, **18**, 229.

—— (1935). Sur la formation d'une rangée axiale de somites chez l'embryon de poulet après irradiation du nœud de Hensen. *C.R. Soc. Biol.* **118**, 452.

—— (1936). Les bases de la tératogenèse expérimentale des Vertébrés amniotes, d'après les résultats de méthodes directes. *Arch. Anat., Strasbourg*, **22**, 1.

—— (1948). La duplication de l'axe embryonnaire et la polyembryonie chez les Vertébrés. *C.R. Soc. Biol.* **142**, 1282.

WOLFF, E. & KAHN, J. (1947a). La regulation de l'ébauche des membres chez les Oiseaux. *C.R. Soc. Biol.* **141**, 915.

268   BIBLIOGRAPHY

WOLFF, E. & KAHN, J. (1947b). Production expérimentale de la polydactyle chez l'embryon d'oiseau. *C.R. Acad. Sci.* **224**, 1583.

WOLFF, E. & LUTZ, H. (1947a). Sur la production expérimentale de jumeaux chez l'embryon d'oiseau. *C.R. Acad. Sci.* **224**, 1301.

—— —— (1947b). Sur une méthode permettant d'obténir expérimentalement le dedoublement des embryons d'oiseaux. *C.R. Soc. Biol.* **141**, 901.

WOLFF, E. & STÉPHAN, F. (1948a). Analyse expérimentale de la détermination des gros vaisseaux extra-embryonaire chez l'Oiseau. *C.R. Soc. Biol.* **142**, 1018.

—— —— (1948b). Sur les méthodes permettant de modifier le développement du systéme vasculaire de l'embryon des Oiseaux. *C.R. Acad. Sci.* **227**, 1270.

WOODSIDE, G. LI. (1937). The influence of host age on induction in the chick blastoderm. *J. Exp. Zool.* **75**, 259.

YNTEMA, C. L. (1944). Experiments on the origin of the sensory ganglia of the facial nerve in the chick. *J. Comp. Neurol.* **81**, 147.

YNTEMA, C. L. & HAMMOND, W. S. (1945). Depletions and abnormalities in the cervical sympathetic system of the chick following extirpation of neural crest. *J. Exp. Zool.* **100**, 137.

—— —— (1947). The development of the autonomic nervous system. *Biol. Rev.* **22**, 344.

ZWILLING, E. (1934). Induction of the olfactory placode by the forebrain in Rana pipiens. *Proc. Soc. Exp. Biol. Med.* **31**, 933.

—— (1942a). Restitution of the tail in the early chick embryo. *J. Exp. Zool.* **91**, 453.

—— (1942b). The development of dominant rumplessness in chick embryos. *Genetics*, **27**, 641.

—— (1945a). The embryogeny of a recessive rumpless condition of chickens. *J. Exp. Zool.* **99**, 79.

—— (1945b). Production of tail abnormalities in chick embryos by transecting the body during the latter part of the second day of incubation. *J. Exp. Zool.* **98**, 237.

—— (1946). Regulation in the chick allantois. *J. Exp. Zool.* **101**, 445.

—— (1948a). Insulin hypoglycemia in chick embryos. *Proc. Soc. Exp. Biol. Med.* **67**, 192.

—— (1948b). Association of hypoglycemia with insulin micromelia in chick embryos. *J. Exp. Zool.* **109**, 197.

—— (1949). The role of epithelial components in the developmental origin of the 'wingless' syndrome of chick embryos. *J. Exp. Zool.* **111**, 175.

ZWILLING, E. & DEBELL, J. T. (1950). Micromelia and growth retardation as independent effects of sulfanilamide in chick embryos. *J. Exp. Zool.* **115**, 59.

INDEX

Printed in the United States
By Bookmasters